"There are many books that discuss why f [illegible] . there are few that move on to discussing how to [illegible] them. *Thriving with Stone Age Minds* is an exciting exception. Justin Barrett and Pamela King are able guides who give us a way to think about evolution in such a way that doesn't flatten the world but opens us to mystery. Beautifully written and richly argued, this book is a must for anyone who is interested in evolution and faith. For pastors and students in practical theology, this book is a treasure."

Andrew Root, Carrie Olson Baalson Professor of Youth and Family Ministry at Luther Seminary, author of *The Congregation in a Secular Age*

"The nature of the flourishing life has been at the heart of debates for millennia. Competing and often incommensurable traditions describe, articulate, and recommend their vision of the good life. For centuries, the question of what it means to flourish was at the heart of great religions and philosophies, and science took a back seat. No longer. Barrett and King's goal is to bridge the historically unbridgeable gulf between Christian theology and evolutionarily psychology. Their rapprochement is intellectually engaging, emotionally satisfying, and spiritually illuminating. This book is required reading for the thinking Christians wrestling with the most important human question: What is the true, flourishing life, and how can we live it?"

Robert Emmons, professor of psychology at University of California, Davis, and author of *The Psychology of Ultimate Concerns*

"The kingdom, church, and world need more writing with this kind of thoughtful critical analysis. *Thriving with Stone Age Minds* is a welcome addition to the canon of faith and science and provides abundant examples for how to incorporate evolutionary psychology into one's daily Christian life."

Matthew Nelson Hill, associate professor of philosophy at Spring Arbor University

"This elegant, engaging, and at times witty book brings evolutionary psychology into constructive dialogue with Christian theology. With exemplary conceptual and logical clarity, it works toward a vision for human thriving that is congruent with both contemporary psychological science and biblical tradition. Full of scholarship but always grounded in the recognizable world of everyday human life, it offers some intriguing possibilities for further reflection and practical action."

Joanna Collicutt, University of Oxford, author of *The Psychology of Christian Character Formation*

"In a world that seems to change at ever-accelerating speed, Justin Barrett and Pamela Ebstyne King help us understand how our evolutionary past gives us particular, ancient tools for negotiating our world and why that's dizzying. At the same time, they demonstrate how our 'Stone Age minds' can be transformed so that we thrive in the telos of the Gospel."

Greg Cootsona, author of *Mere Science and Christian Faith*, faculty in religion at Chico State University

"This book articulates in a fresh way why and how purpose and other virtues are essential to human flourishing. Educators, parents, and others concerned about our public and private thriving need to fully understand these critically important ideas. Barrett and King's account expands the reach of these ideas through a close look at the intersection of three frames of reference: moral psychology, evolutionary biology, and Christian thought. The result is unique and fascinating."
Anne Colby, coauthor of *The Power of Ideals*

"This book is a model of Christian discernment, considering the latest scientific research in the light of Christian faith to address challenging questions of our day. Barrett and King explain the established findings of evolutionary psychology (EP) while shedding the pop psychology and anti-God baggage that is often added to it. Yet their goal is not merely showing the intellectual compatibility of EP with Christianity but 'placing evolutionary psychology in the service of Christian theology' to address larger questions. By better understanding human nature and the traits that make us unique from animals, we can better fulfill our purpose as God's image bearers and work more effectively to build communities that thrive in the ways God intended. The book's readable style, vivid examples, and helpful study guide make it a great book for students, book clubs, science fans, and pastors."
Deborah Haarsma, president of BioLogos

THRIVING WITH STONE AGE MINDS

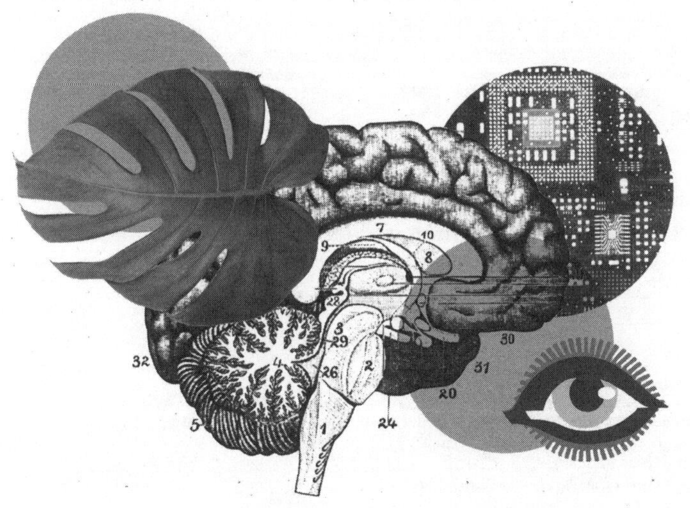

Evolutionary Psychology, Christian Faith, *and the* Quest for Human Flourishing

JUSTIN L. BARRETT *with* PAMELA EBSTYNE KING

InterVarsity Press
P.O. Box 1400, Downers Grove, IL 60515-1426
ivpress.com
email@ivpress.com

InterVarsity Press® is the book-publishing division of InterVarsity Christian Fellowship/USA®, a movement of students and faculty active on campus at hundreds of universities, colleges, and schools of nursing in the United States of America, and a member movement of the International Fellowship of Evangelical Students. For information about local and regional activities, visit intervarsity.org.

Cover design and image composite: David Fassett
Interior design: Jeanna Wiggins
Images: computer motherboard: © Peter Dazeley / The Image Bank / Getty Images
paper background: © Peter Dazeley / The Image Bank / Getty Images
brain illustration: © dani3315 / iStock / Getty Images Plus
large green leaves: © sarayut Thaneerat / Moment Collection / Getty Images
eye illustration: © CSA Images / Getty Images

ISBN 978-0-8308-5293-2 (print)
ISBN 978-0-8308-8849-8 (digital)

Printed in the United States of America ♾

InterVarsity Press is committed to ecological stewardship and to the conservation of natural resources in all our operations. This book was printed using sustainably sourced paper.

Library of Congress Cataloging-in-Publication Data
Names: Barret, Justin L., 1971- author. | King, Pamela Ebstyne, 1968- author.
Title: Thriving with Stone Age minds / Justin L. Barret with Pamela Ebstyne King.
Description: Downers Grove, IL : InterVarsity Press, [2021] | Series: Biologos books on science and Christianity | Includes bibliographical references and index.
Identifiers: LCCN 2021009291 (print) | LCCN 2021009292 (ebook) | ISBN 9780830852932 (print) | ISBN 9780830888498 (digital)
Subjects: LCSH: Psychology and religion. | Evolutionary psychology. | Theology. | Theological anthropology—Christianity.
Classification: LCC BF51 .B37 2021 (print) | LCC BF51 (ebook) | DDC 200.1/9—dc23
LC record available at https://lccn.loc.gov/2021009291
LC ebook record available at https://lccn.loc.gov/2021009292

P 25 24 23 22 21 20 19 18 17 16 15 14 13 12 11 10 9 8 7 6 5 4 3 2 1
Y 37 36 35 34 33 32 31 30 29 28 27 26 25 24 23 22 21

For Jeffrey Schloss,

Someone from whom I have learned a lot concerning how to think about evolution Christianly and also the person who should have written this book but could not be persuaded to do so.

JLB

For Aidan, Rhys, Jocelyn, Max, Anika,

Jones, Callie, Hannah, Elyse, Jessica, and Nik

The next generation of my family who give me great hope in the human species's ability to thrive through the complexity of life and who actively and uniquely contribute to the flourishing of our earthly niche.

PEK

CONTENTS

INTRODUCTION

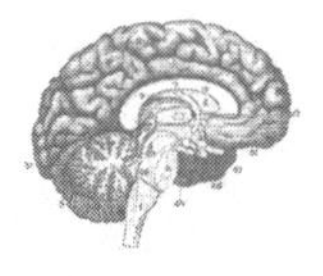

Something isn't right. Sure, globally, life expectancy is rising and infant mortality is falling. Violence—including warfare—is lower in the twenty-first century than the twentieth century. Education is up and starvation is down. By some estimates, less than 10 percent of the world's population now lives in "extreme poverty," while over 70 percent did in 1900.[1] There is a lot to be grateful for, and yet . . .

In many nations with high levels of economic development, including the United States, people are taking longer to gain financial independence and start their own families. Depression, suicide, and drug addiction continue to ravage lives, particularly the lives of young people. New addictions and addiction-like dependencies—to social media, video gaming, gambling, and pornography—keep popping up. Anxiety disorders may be more common than ever before.[2] Adequate food, security, and other basic resources for living are available to an increasingly large share of humanity, but we don't seem to be flourishing. Something isn't right.

This book is concerned with integrating the best of the human sciences with the best of theological insights around the topic of human

[1]Roser & Ortiz-Ospina, 2019.
[2]Renner, 2018.

flourishing or thriving. Specifically, I suggest that evolutionary psychology and neighboring scientific disciplines can serve as useful tools in analyzing the challenges to human thriving and possible solutions.[3] These tools cannot create full-bodied solutions on their own, but joined with philosophical and theological wisdom, they may be tremendously valuable. Because of this value, this book serves as an introduction to evolutionary psychology contextualized around a particular problem: what does it mean to thrive, and why does it often seem so difficult?

GIVING EVOLUTIONARY PSYCHOLOGY A CHANCE, GIVING CHRISTIAN THEOLOGY A CHANCE

I am not here to take sides. I am not on the side of evolution or the side of biblical Christianity, or even on the side of the idea of "sides." My aim is to pursue understanding and truth. That aim, however, starts with some suppositions. I am convinced that the sciences, at their best, are powerful tools for revealing some truths, but not all. The relevant scientific work currently converges on the idea that we humans have come to be the creatures we are in part through a process of evolution from other species. I am also convinced that the Bible, properly interpreted, reveals critical truths about human nature and purpose and what it means to live a life worth living. The Bible reveals that humans are creatures; that is, they are created beings and created in God's image. The sciences derive much of their explanatory power from the fact (as I see it) that God created an orderly world and made at least some organisms, human

[3]Though this book rose out of a collaboration among Drs. Pamela King, James Furrow, Sarah Schnitker, William Whitney, Tyler Greenway, and myself, I (Justin Barrett) am the primary author and stand behind all of the claims (at least right now). For this reason, and so as not to make for confusing prose, I use the first-person singular pronoun throughout. Nonetheless, there are several sections where Pam King took the lead on writing, including large sections of chapters seven and eight (particularly concerning human telos), as well as the discussion of purpose in chapter three and the material on mentoring and sparks in chapter five. She has made numerous valuable contributions throughout concerning contemporary scholarship on positive youth development, thriving, and theology. These contributions exceed ordinary collegial courtesy and merit attribution through authorship, but she does not necessarily stand behind all the arguments the way I do. Though she has attempted to keep me from grave errors and we stand together behind much of this text, she should not be burdened to defend it in its entirety.

beings, capable of partially and imperfectly glimpsing that order. Importantly, the Bible also reveals that, though some truths can be apprehended intellectually, ultimate truth was embodied by a person: Jesus of Nazareth, wholly human and wholly divine. What it means to perfectly image God and live an abundant, thriving life was captured in the life of Jesus. These are some of my starting points, but you need not share them to profit from this book.

I am also convinced that when it comes to science, biblical interpretation, and theological reflection, we are bound to get some things wrong. History shows us repeatedly how scientific theories have had to be amended or even discarded. The scientific consensus of one generation passes away, but not usually in its entirety. In psychological science, it was not long ago that many prominent scientists regarded nearly all human behaviors as the product of fairly simple learning mechanisms. As I will explain, psychological science has moved on from this position. We haven't rejected the importance of basic learning mechanisms, but we have added to and refined the story. Scientific inquiry has shown progress. Less dramatically, biblical interpretation and theological reflections on Scripture have also shown progress. For instance, the idea that God is a unity of three "persons," Father, Son, and Holy Spirit, is not clearly and unambiguously articulated in the Bible but took centuries of theological development after the Bible was written. We even see some change of thinking concerning God's revelation within the Bible itself. Jesus' earliest followers, the apostles, initially assumed that any followers of Jesus, even non-Jews, had to adopt Jewish dietary restrictions. We read in the book of Acts how the apostle Peter changes his mind on this matter (see Acts 10). Theological progress happens. For these reasons, I have written this book with the assumption that I have gotten some things wrong and I will eventually change my mind.

I intend this book for two primary audiences. The first is people who are convinced of the authority of the Bible (properly interpreted) but not convinced that God created humans through a process of evolution from other organisms. If you are in this audience, I ask you to suspend your discomfort with evolution for a bit. I am not going to lay out a case for human evolution; many other books have done that.[4] Rather, suppose God did choose to use evolution as a means for making human beings. If this turns out to be true, how could that perspective change how we think about human thriving? Does evolutionary psychology specifically give our theological considerations new support or insights?

My second audience is people who are convinced that humans have evolved from other species but are not confident that this understanding coheres with the idea that the Bible has valuable insights concerning what it means to live a thriving life. Perhaps you worry that you have to pick sides between evolutionary science and biblical authority. If you are in this audience, I ask you to consider how much more powerful evolutionary psychology is in considering human purpose and the good life when infused with biblical, theological perspectives. Maybe you'll find that your doubts about Christian theology are not as serious as you once thought.

Both audiences may wonder how Christian theology and scientific inquiry concerning human thriving can be brought together. It is common to assume, for instance, that the whole point of faith is that you don't need evidence for it. Faith is believing without evidence, right? Not really. The Christian faith is an evidential one. It hangs on the claim that a real person who really existed led a sinless life and was really executed but really came back from the dead. If that did not really happen, one might still have reason to believe in some kind of God, but Christianity would be false and should be abandoned.[5] The point of New Testament books such as the Gospel according to Luke and the Gospel according to John is to give testimony and evidence in support of the reality that Jesus was "God with us," the Savior of all humanity.

---

[4]For some personal accounts of how various theologians, philosophers, and scientists who are Christ-followers became persuaded that evolution is the best available account of how God brought about humans, see Applegate and Stump (2016).

[5]Consider 1 Corinthians 15:12-19, where Paul writes that if Jesus Christ did not actually rise from the dead then "we are of all people most to be pitied."

So a Christian worldview requires evidence but not perfect and comprehensive evidence (if that were even possible). It needs enough evidence to build our faith but not so much as to be a substitute for it. Not all of the evidence is of the typical scientific sort, directly available to the senses of multiple observers and measurable. Some evidence will be personal experience, testimonial, historical, and logical. Nonetheless, the sciences are a sensible place for even Christians to turn for evidence of the elements of a thriving human life. The sciences are tools for seeking understanding that have proven their usefulness over the centuries. It is not unreasonable to suppose that a scientific approach to human thriving will bear fruit too.

But it may seem that the question is too value-laden for a scientific approach to be of much help. After all, the sciences describe what things are and how they work, why X followed from Y, but they don't tell us what we should and shouldn't do. The sciences, including evolutionary psychology, can help us see how things are or have been but not how they ought to be, at least not by themselves. The sciences need a healthy complement of additional philosophical or theological assumptions and considerations before one can conclude, "Therefore, we should . . . " For this reason, my project is not to leave evolutionary psychology to stand on its own but to consider what evolutionary psychology might teach us, then to combine those insights with additional resources from Christian theology.

WHY EVOLUTIONARY PSYCHOLOGY?

Let me lay bare the logic of this book for each of my two audiences. First, if you are coming to this book with traditional Christian commitments, here is the rationale for our focus on evolutionary psychology:

- Humanity has been created by God in His image.
- All of humanity, not just one or two individual humans, was created in God's image, and we know of no other living things that are also God's image bearers.

- God was not arbitrary in designating humans—but not other animals—as created in His image.
- Therefore, living well as God's image bearers has something to do with what makes humans distinct from other animals and how that distinctively human nature may be lived out in various environments.
- The sciences that have studied human nature are helpful tools for understanding human nature and how to effectively pursue our created purposes, especially in comparison to other animals—considering both the similarities and the differences.
- Evolutionary psychology is one key science that has studied the properties that make humans unique and why, as well as how, human nature interacts with environments.
- Therefore, it may be useful for Christians to examine what evolutionary psychologists have discovered concerning human nature in order to more fully articulate a vision of human thriving.

Perhaps you are not coming with particularly strong Christian commitments but are more enthusiastic about evolutionary sciences. Then one way to understand the rationale of the book is this:

- Thinkers through the ages have commonly considered at least three dimensions of human flourishing or thriving: a life that is happy or feels good; a life that brings about good things, is harmonious, or goes well in relation to other people and one's context; and a life that is lived well based on certain ethical or moral principles.[6]
- Regardless which of these three dimensions of thriving you wish to emphasize, evolutionary psychology may make a contribution.
  - If one stresses happiness, evolutionary psychology can help us see why some things generally make us happy or are otherwise attractive to us. For instance, happiness responses may have

[6]See Volf and Croasmun (2019) for an overview of human flourishing (or thriving) using these three focuses.

motivated our ancestors to act in ways that improved their fitness. If so, evolutionary psychology may enable us to better understand the contexts in which we are likely to maximize happiness.

- If one stresses harmony with our situations and surroundings, evolutionary psychology may help us see how our species has found distinctively human solutions to environmental (including social) demands such that, on average, our ancestors led good enough lives to have strong fitness. Indeed, the concept of fitness builds in it the notion of fitting well with the demands in one's life.
- If one's approach to thriving emphasizes living life according to principles about what is right and ethical, it is almost certain that part of those considerations include how they impact the self, other people, and the environment. That is, the consequences of those principles will usually play some role in determining whether they guide a good life or not. Here, too, evolutionary psychology may help by providing evidence and theories that concern the consequences of certain thoughts and actions that humans find themselves drawn toward.

- In all three of these dimensions, some intellectual work has to be undertaken to determine the relationship between thriving and the evolutionary concept of fitness. Nevertheless, exploring the exact nature of the relationship between these two concepts is bound to sharpen one's thinking on what makes for a thriving life.

From either perspective, an exploration of evolutionary psychology and what it can tell us about human nature (and how that nature may have developed in relation to various environmental challenges) will help hone our thinking about what it means for humans to thrive. Consequently, much of this book summarizes what we can learn about human nature via evolutionary theory, especially evolutionary psychology. Only after this summary do I turn to how these insights may inform a Christian understanding of human thriving.

To sum, this book serves as an introduction to evolutionary psychology by way of exploring human thriving. Though my primary aim is to demonstrate the potential utility of bringing theology and evolutionary psychology together to address this and other important topics, my hope is that you will also be provoked to think anew about what it means to thrive in your own life and in the communities of which you are part.

1

WRESTLING WITH EVOLUTIONARY PSYCHOLOGY, EMBRACING CHRISTIAN THEOLOGY

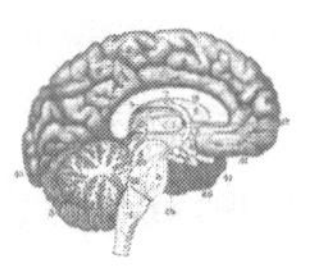

It took me a long time to get comfortable with the idea of human evolution, let alone evolutionary psychology. I still find myself wincing when people label me an "evolutionary psychologist." That is not how I identify myself. I am a scholar with training in psychological science, and when evolutionary perspectives are helpful in addressing a particular problem, I try to use them judiciously. But even this measured degree of comfort with evolution and evolutionary psychology has been slow to develop, and I don't begrudge those who are not sure what they think.

I grew up in a family that took the Bible's ability to authoritatively provide insight and direction for our lives very seriously. It was common in my family to talk about what the pastor had just said about this part of the Bible and whether it fit with that other part of the Bible and whether his interpretation was on target. I was also taught early on that the Bible clearly conveys a young earth that was spoken into existence in six literal, twenty-four-hour days several thousand years ago (I even did the calculations myself from biblical genealogies!) and that human evolution was a sinister conspiracy meant to undermine the Bible and God's sovereignty.

I still remember learning a song through a church program that included the lines, "I'm no kin to the monkey. The monkey's no kin to me. I don't know about your ancestors but mine didn't swing from a tree." I didn't become comfortable with evolutionary theory until graduate school.

What made the change? In part, I learned a bit about biblical interpretation through undergraduate courses at Calvin College (now Calvin University). I learned that the churches I grew up in were not merely letting the Scripture speak for itself but were interpreting its words through the lens of a particular tradition that not all sincere and thoughtful Christ-followers shared. Maybe there were other legitimate ways to understand what God meant for us to learn about Him[1] and ourselves through the two creation accounts in the book of Genesis and other key passages in the Bible (e.g., Job and the Psalms).

I also discovered that I was approaching evolutionary theory from a position of fear: I was afraid of what it might mean for certain theological convictions I held. A zoology professor, while compassionately affirming the concerns shared by those of us from more conservative backgrounds, helped us see that if God said He made humans in His image, then He created us in His image, whether He used a process of evolution or a special single action. And even if humans evolved gradually, this did not change the fact that at some point they were in a position to have a relationship with God but also able to reject that relationship and become subject to sin—a condition that continues to plague humanity and from which we need God to rescue us. That is, I came to see that my fears were giving evolutionary theory more power to demolish Christian theology than it deserved to have.

Even when I became comfortable with human evolution, I was not so sure about evolutionary psychology. At first acquaintance, evolutionary psychology can look suspect. Its most sensational studies seem to

[1]I break with current fashion and use capitalized pronouns for God (the Father or Holy Spirit) in order to remind readers—and myself—that God is not literally male in the way that a human being is. This is a metaphorical personal pronoun. I do not, however, use capitalized masculine pronouns for Jesus as he was male in his humanness.

simplify human behavior to strategies for making babies. It is easy to gloss evolutionary psychology as a reduction of human behavior to genes or ignoble instincts, a deflation of the differences between us and other animals, or an inflation of the differences between people of different ancestral lineages. And if male and female behavior really is all about making babies, if we really are just vehicles driven around by particular genes, if our evolved genes are the real story, or if genes cause human group differences, then doesn't evolutionary psychology provide intellectual cover for sexism and racism?

No. First impressions can be misleading. Evolutionary psychology, particularly as it has matured in the last decade, is not fairly characterized as sex-crazed or fixated on humans as nothing but genes, and it certainly cannot be easily used to support devaluing others on the basis of their sex or race. But I understand if you associate evolutionary psychology with these unsavory ideas, as I did. Sex and controversy sell, and they have been used to sell evolutionary psychology.

Evolutionary psychology is the scientific study of human thought (including feelings), relationships, and behaviors that grounds its work in the claim that humans living today are the way they are in large part because of past selective pressures working on human ancestors. That is, humans have evolved from previous conditions and even previous species and bear the marks of that past. From that perspective, many otherwise puzzling or inexplicable features of human thought, relationships, and behaviors become more understandable.

Humans have evolved from previous conditions and even previous species and bear the marks of that past.

For instance, we may wonder why humans seem to be so strongly attracted to doughnuts. Why do doughnuts taste so good to most of us even though too many can make us feel unwell and, over time, damage our health? If it is true that humans have the tastes they do to solve problems that our ancestors faced, it could be that our attraction to things that taste sweet and have a high fat content evolved at a time when obtaining sweet or fatty things was hard work and intermittent but carried an important nutritional payoff. Sweet things like fruits are high in valuable vitamins, but picking berries can be laborious and lead to lots of cuts and scratches (depending on the type of berry). Berries and fruits are also targeted by many other animals and can spoil quickly. Fatty things, like animal fats or nuts, can require a lot of work to obtain and lots of preparation to eat safely and may be available only sporadically, but they may be necessary for proper brain development. So our ancestors needed strong motivation to put in the hard work or face the risks necessary to obtain these kinds of foods and, perhaps, eat a lot in short periods of time.

Those of our ancestors who liked sweet and fatty things better than their peers were more motivated and more rewarded and, as a result, survived and reproduced better than their peers who didn't care about sweet or fatty foods. Gradually, our ancestors' population as a whole became more attracted to sweets and fats. We are descended from that population. If this account is (roughly) right, we have a better understanding of why doughnuts taste so yummy to us and why they may be so hard to resist. Doughnuts exploit our nature. Of course, offering a plausible-sounding story isn't enough. The science requires gathering evidence that corroborates or challenges such an account and rules out alternative competing accounts, and that is what evolutionary psychologists and their colleagues in neighboring scientific disciplines do.

As a subdiscipline that considers how humans (and human ancestors) have interacted with their environments over eons, evolutionary psychology may be especially well-suited for considering what it means for humans to thrive or flourish. Evolutionary perspectives can help us see what is commonly taken as evidence of thriving in any species, such as growing up, staying alive, and having babies, and also how humans seem to solve problems in characteristically human ways. If we want an account of human thriving, it may be helpful to look at how humans are distinct from other animals. Evolutionary psychology can help us make such comparisons.

Nevertheless, thriving cannot simply be collapsed to long lives and successful reproduction. These properties may be indicators of what evolutionary scientists call fitness, but fitness is not the same as thriving. Fitness is the degree to which an organism can successfully survive and reproduce in its particular environment. It is not hard to imagine a future scenario in which humans have created societies of human clones that serve as slave labor for a small ruling class but that also live a long time and continue to grow in population. This would be a society marked by fitness but not a society of thriving people.

Similarly, thanks to human selective breeding and agribusiness systems, the descendants of Southeast Asian jungle fowl—what we call chickens—now are more numerous than they could ever have been left to their own devices. But are chickens in barns in a greater state of thriving than jungle fowl? Perhaps some are, but it isn't so obvious. Fitness is not the same as thriving.

I want to reserve the concept of thriving as a synonym for flourishing, suggesting some kind of good and full life, a life well lived and properly enjoyed, a life that has a positive impact on its surrounding.[2] Whereas fitness is a concept scientists use descriptively concerning what is or has been the case, I want to use thriving as a prescriptive concept of what should or ought to be the case. To do so, I must add to the science.

SOME CHRISTIAN THEOLOGY TO GET US STARTED

Conventional Christian theology converges on a few ideas that will factor importantly in my attempt to integrate evolutionary psychology and theology around the notion of human thriving.

Imago Dei. First, humans are created by God in God's image (*imago Dei*). According to tradition, this state of being created in God's image is not shared with any other animal on earth, so humans are unique in this regard. Thus theologians commonly treat the ideas of *imago Dei* and human uniqueness as closely related. Living into this

[2]See King and Mangan (in press).

status as *imago Dei* and recognizing it as one's purpose or telos may be key to thriving as a human.[3] Because the scientific study of human properties in comparison with those of other existing animals has much to say about human uniqueness, unsurprisingly, theologians have begun considering human uniqueness and the *imago Dei* in connection with relevant sciences.[4]

The perfect humanness of Jesus. The Christian church almost universally affirms that Jesus Christ was fully divine but also fully human. Indeed, Jesus most perfectly lived out what it means to image God. Jesus most perfectly lived out his life's purpose and best demonstrated what it means to thrive. Furthermore, humans are called to use him as an example for how to live. For instance, the apostle Paul encouraged the Corinthians to imitate him (Paul) insofar as he was imitating Jesus (1 Corinthians 11:1). Likewise, in Paul's letter to the Ephesians, he calls on his readers to imitate God and provides Jesus' example to imitate (Ephesians 5:1-2). This theme appears repeatedly throughout the New Testament (e.g., 1 John 2; 1 Peter 2). Thus an important part of what it means to thrive—to become what God created us to become—is to imitate Jesus' life in relevant ways. Jesus is a guidepost for thriving.

Jesus most perfectly lived out his life's purpose and best demonstrated what it means to thrive.

The help of community: the church. Jesus surrounded himself with a community of followers who were attempting to imitate him in loving God and loving others. These followers became the foundation of the church. The global church refers to the worldwide community of Christ-followers, but it is also visible in a local community of Christ-followers. The Bible refers to the universal or global church as both the body of Christ (1 Corinthians 12) and the bride of Christ (2 Corinthians 11;

[3]For a teleological approach of this kind, see King (2016, 2020) and King and Whitney (2015).
[4]For instance, see Jeeves (2015); Rosenberg, Burdett, Lloyd, and van den Torren (2018); and van Huyssteen (2006).

Ephesians 5; Revelation 19), two metaphors emphasizing the intimacy the church should have with Jesus. Like its individual members, the church as a whole and any given local church should conform to Jesus' purposes and resemble him. I argue that a church thrives as much as it promotes the thriving of those who compose it and collectively promote the mission the church has been given by Jesus: advancing God's kingdom.

The mission of the church: advancing God's kingdom. A pivotal component of Christian theology is the idea that God deliberately, intentionally created the world. It is God's world first and foremost. God has, however, given the world—particularly humanity—freedom to align with God's kingship or reject God's rightful rule and direction for the world. In big and little ways humans have chosen their own direction, their own sense of what is good and right, over God's. God has not, however, given up on the world. God pursues us to restore us to our proper place in the created order and to bring all of creation back into alignment under divine rule, a creation characterized by love, service, and care. Jesus taught that he came to serve God in advancing God's kingdom, and it is the role of his followers, the church, to do likewise. That is, the church is to advance God's kingdom. That is its purpose or telos. The church thrives by advancing the kingdom of God to include all of creation.

THE THESIS IN BRIEF

With these theological points in mind, I can offer a brief synopsis of my view of what it means for humans to thrive. Human thriving is becoming what God has created us each to become, to live into our purpose(s).[5] God has created us as *imago Dei*, as bearers of God's image, and so the more closely we image God, the more we thrive.[6] God created and selected humans to be His image bearers—to reflect Him in the world—in part because of special features that humans alone have the potential to

[5]See King (2016, 2020) and King and Whitney (2015).

[6]If we take sanctification—becoming holy—as, at least in part, a process of becoming more Christlike, then thriving and sanctification become comparable concepts (see King & Argue, 2020).

actualize. Evolutionary psychology helps us see what human nature is, which helps us see what distinctively human thriving is. Humans have three interacting clusters of distinctive capacities: being very social in some interesting ways, having an unparalleled capacity to acquire and share information with each other, and possessing an unusual ability to willfully control themselves. These three groups of features make humans the only animals alive today that can image God by exercising His reign over the rest of creation and, most importantly, the only animals that can truly and deeply love God and love each other.

Jesus showed us how this imaging is done better than anyone. In Colossians 1:15 we read that Jesus "is the image of the invisible God," and Hebrews 1:3 asserts that Jesus "reflects the glory of God and bears the very stamp of his nature." Thus, the more we imitate Jesus (in the relevant respects), the more we image God well and we thrive. Jesus established a community of followers—the church—to work together in individually imitating him and collectively participating in the kingdom of God. So as social beings we are aided in our thriving by being part of a thriving local community of Christ-followers that is part of a thriving global church. The church, locally or globally, is thriving when it is becoming what God created it to be: a community that equips its members in advancing the kingdom of God such that all of creation will become what God created it to be. Thriving, then, from this theological perspective is a series of rings, like a bull's-eye: the individual imitating Christ at the center surrounded by various sizes of communities. The more closely our various communities of relationships resemble what a Jesus-oriented community is supposed to be, the easier it will be for us to each hit the target as individuals and thrive. Our potential for thriving increases, then, when the communities we are part of also move closer to Jesus and center on his example.

This view of human thriving is directional.[7] To thrive is to move toward becoming the person you were created to be, to approach who it is God

[7]King, 2018; King & Whitney, 2015.

wants you to be. But this simple formulation hides some important complexity. Because our thriving is not isolated from the world around us and because our thriving is, in part, to love others well, our thriving must be reciprocal with others and generative. That is, we thrive when we help others around us thrive, when we improve the conditions for others' thriving. How we naturally fit into the environment around us—social and otherwise—is part of the recipe for thriving.

To thrive is to move toward becoming the person you were created to be, to approach who it is God wants you to be.

Evolutionary psychology teaches us an important lesson that informs these claims: one of the distinctive features of humans is the degree to which we solve fitness problems by changing our environment around us, our niche. Like other animals, we humans change our behavior to meet the demands of our niche, but to an extent that is unparalleled we also change our niche to make it more accommodating to our desired behaviors.

And yet our remarkable ability to change our environments to solve today's fitness challenges can create new problems for us and our offspring for which our nature may not be terribly well-suited. For instance, we humans have formed cities teeming with machines to make it easier to cooperate with each other to share food, medicine, housing, and other good things. But one byproduct is noise. The World Health Organization has linked environmental noise exposure to heart disease, sleep disturbance, cognitive impairments, irritability, attention difficulties, and other mental health risks. These links may be largely correlational (and therefore could be caused by factors other than noise), but many of them plausibly follow from being constantly bombarded by noise levels our physiology and psychology are not naturally equipped to handle.[8]

Food allergies may represent a similar problem. Food allergies are on the rise, especially in industrialized or postindustrialized societies. Why

[8]World Health Organization, 2011.

would we suddenly—through no genetic change—begin to form potentially deadly allergies to our foods?[9] A leading theory, which I was told by my allergist some years ago, is that the part of our immune system that used to handle invading parasites such as tiny roundworms now has too little to do because of the abundance of clean water and our removal from natural ecosystems. So we have managed to address one fitness problem—parasitic disease—but the result is that our natural mechanisms for fighting parasites mistake peanuts, eggs, and wheat for dangerous intruders. Our nature mis-fits with our new environments, so now we have a new problem to solve in order to thrive.

The trick seems to be, how can we thrive in the environment we live in without changing it so much that those who come after us struggle to thrive? Imagine a group of children at a lake, standing on a dock. A little way from the dock is a large inflatable raft. The game is to jump to the raft. You watch as the boldest kids jump from the dock onto the raft one by one. But then you notice something. With each jump, the raft drifts a little farther away from the dock, making it harder for the next child to make the jump. Soon, children can't make the jump anymore but land in the water just before the raft and have to clamber up onto it, pushing it still farther away, making it even harder for the subsequent children to reach it.

A particular challenge for humans, because of the kind of animal we are, is not to let our pursuit of thriving be like this game. Our thriving should not place undue burdens on others; that is not loving others well. How, then, do we traverse the gap—the gap between our nature and the demands of our environment—such that we do not make the gap wider for others? Or have we already created such a huge gap between our nature and our environment over the ages that we need to narrow it to facilitate thriving? A central point of this book is to demonstrate that placing evolutionary psychology in the service of Christian theology is a promising strategy for making progress in answering these questions.

---

[9]Santos, 2019. Like all science, medical science is always changing, and this may not prove to be the best account of all allergies, but it is illustrative of the basic point: ills today may be due to solutions for past environmental demands.

The idea of using evolutionary psychology as a tool for theological inquiry may strike you as peculiar or even outrageous. I ask you to give me a chance to make the case. Much of the heat around evolutionary psychology has arisen not because of the science itself but because of extra baggage that some scholars have packed in with it, baggage I am happy to leave behind. As in every new and rapidly growing intellectual area, evolutionary psychology is checkered with missteps and excesses. Nevertheless, behind the occasional grandstanding of its advocates, there are some basic assumptions that should resonate with Bible-affirming Christians. Evolutionary psychology, like traditional Christianity, emphasizes an ancient common human nature that binds us all together regardless of ethnicity, nationality, or social class. It also resonates with creation theology's insistence that humans are creatures and not gods; we share a basic kinship with other creatures under God's sovereignty. We, too, are dust of the earth, but fearfully and wonderfully made. Furthermore, both Christian theology and evolutionary psychology have spent a fair amount of time exploring what makes humans distinctive or even unique among the creatures. As I will explain further, we see remarkable convergence on the details of human distinctiveness in evolutionary psychology and Christianity. My aim, however, is not to argue for the harmony between evolutionary psychology and a biblical Christian faith, but to demonstrate it by showing the utility in bringing the two together around a topic that should concern us all: What does it mean to thrive?

Humans are attempting to thrive in contemporary contexts with Stone Age minds. The speed with which our ancestors changed the world was greater than the speed with which our psychology evolved to keep up with those changes. This mismatch between our psychological natures and the environments in which many of us find ourselves is a big part of why thriving seems so hard even if surviving is easy. And we have not stopped creating gaps. We are continuing to change our environments and, thus, perpetually having to mind the gap.

2

WHAT IS THRIVING?

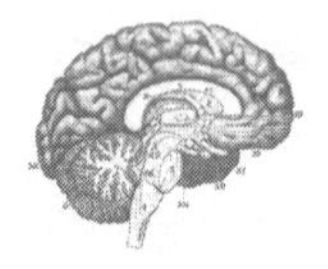

I confess that I am suspicious of people who try to sell me things, even when they are health care providers. When a long string of dentists over the years suggested that I ought to have braces to straighten out my crowded teeth, I politely declined, thinking someone else could make their boat payments. In more recent years when a dentist would ask if I had considered braces or newer technologies for my teeth, I asked whether it was medically necessary or just cosmetic. I could live with an ugly mouth if it worked fine. Last year, a dentist finally persuaded me that I needed work on my mouth before my teeth started fracturing or I developed problems with my gums. So, here I am, a middle-aged man wearing braces on his teeth. I have joined what is rapidly becoming a majority of Americans who need orthodontics. But that raises a question: Why would even a large minority of people need something to fix a part of us that is so critical for basic survival? Shouldn't we naturally grow teeth and jaws that work properly? Our ancestors didn't need braces, so why do we?

In researching for and writing this book, not only did I acquire braces; I also found a possible answer to these questions. In their book *Jaws: The Story of a Hidden Epidemic*, orthodontist Sandra Kahn and biologist Paul Ehrlich make the case that a number of cultural changes, especially since the industrial revolution, have led to a weakening and shrinking of human jaws, leading to crowded and unhealthy teeth in an increasing number of people. Our hunter-gatherer ancestors rarely had impacted wisdom teeth that needed yanked or crowded front teeth. Why? The factors are numerous and interacting, but one key is the relatively soft, easy-to-eat processed food that now makes up most of our diets, particularly in early childhood. We no longer have to grow up carefully chewing and swallowing tough meats, fibrous vegetables, nuts, seeds, and grains. Sophisticated tools, machines, and cooking devices have given our jaws a rest, so they grow up weak and narrow. Our genes haven't changed; the cultural environment we grow up in has changed, and that has led to biological differences that have implications on our health.

This dynamic captures a central point in this book. Human efforts to address basic challenges of surviving and thriving sometimes change the context in which subsequent generations have to figure out how to survive and thrive. The changes to our cultural or physical environment can be great enough to create new, unintended challenges for the humans who come after us, particularly challenges during childhood. One generation's innovation—like soft foods—creates unforeseen obstacles for the thriving of subsequent generations, such as the need for artificial and expensive gadgets in the mouth to move the teeth and jaws where they would have been had another problem not been fixed. We see this pattern again and again, and that is why, in spite of massive improvements in terms of human well-being on some measures, we still face a number of enormous challenges. Solving one obstacle for thriving eventually creates a new one. A robust sense of thriving, then, includes a sense of an individual's current condition but also recognizes how humans interact with a particular context.

Thrive and *thriving* have become popular buzzwords in contemporary American culture. For instance, at this writing, the medical group Kaiser Permanente uses "Thrive" as the tagline for its medical services. Similarly, my church offers "Thrive" programs aimed at helping people with addiction and similar obstacles that hinder them from becoming who God has created them to be. Physical and psychological health may be an aspect of thriving, but is that all there is to having a full, rich, and abundant life? Is being healthy all there is to thriving? If that were the case, we would expect the dramatic improvement in medical care and nutrition over the past century to be accompanied by a corresponding improvement in human thriving. However, problems with suicide, violence, cheating, anxiety, and related disorders are on the rise, and there does not seem to be a dramatic increase in individuals' sense of well-being. As good as it is to be physically and psychologically healthy, I believe that thriving is more than being well or being less miserable. Our life's purpose is not just to be healthy.

In fact, perhaps you have known someone who was not remarkably healthy but seemed to be thriving. To take one example, I once knew a man who was middle-aged, overweight, unfit, balding, not wealthy, single, slightly deformed, blind, and thriving. He delighted in life and his interactions with others, and he brought tremendous joy to others through his genuine care for them and his gratitude to God for life's small wonders. Thriving is not simply being well or having no difficulties. Sometimes people thrive even if they're diseased or disordered because they use those challenges as a step toward becoming who they are fully meant to be.

Thriving is not simply being well or having no difficulties.

What, then, is thriving? I suggest that, at least in part, thriving concerns a process of becoming who one is meant to be. Perhaps we can gain additional insight by considering nonhuman cases that appear analogous to human thriving.

Have you ever seen a Siberian husky run on a bright, cool day? You can practically see the joy on its face as it does what it was bred to do. The husky is realizing its purpose for life—what it is meant to be. The husky's telos (Greek for "direction" or "purpose")—the reason humans have selectively bred them—is to be a helper for humans. Specifically, they are bred to run, pull sleds, and protect and warm their humans in cool climates. Contrast that image with a husky on a hot summer day in Southern California as it is forced to walk slowly beside its human caretaker. Its countenance is slightly bowed and deflated, its mouth agape, tongue lying limply out one side of its mouth. This contrast suggests that an individual husky shows signs of thriving when it can run and pull and be near its human in a setting appropriate to its physiology. These features are part of the individual dimension of the husky's thriving.

As pack animals, huskies also have a collective dimension to thriving. Huskies are meant to be part of a team of dogs that collaborate to pull a sled swiftly and surely, keeping their humans and each other safe in the process. The whole team can be characterized as thriving when it is functioning the way it is supposed to in order to fulfill its end purpose. A single husky that runs, pulls, and protects its human well is showing individual thriving, but it may or may not be part of a dog team and thus may or may not be part of the team's thriving. In a similar way, humans, as intensely social animals created to be in community with each other, have both individual and collective or communal dimensions to their thriving.

As the case of the husky suggests, my treatment of thriving begins with the assumption that it is a dynamic expression of becoming what one is supposed to be, fulfilling one's purpose. A big part of my inspiration comes from Christian theology, which affirms that we humans, individually and collectively, have a purpose for our existence. Unlike, say, the carefully cultivated Christmas tree that can once and for all fulfill its purpose as a beautiful holiday decoration, humans have much more rich and complex purposes that cannot always be completely fulfilled in this lifetime. I affirm, then, that a thriving, abundant life is characterized by the movement toward and approximation of fulfilling our purpose, of moving toward who God created us to be.[1]

---

[1]King, 2020; King & Whitney, 2015.

In addition to individual and collective dimensions, thriving also has "right-now" and "not-yet-but-becoming" dimensions. Even though Christmas trees and huskies can be said to be thriving because of how well they approximate fulfilling their purposes at any given moment, development is still required for them to fully realize their purpose in this life. A four-month-old husky puppy can show natural potential for becoming a good sled dog, eager to run as part of a pack and eager to please its human, but its limited size, strength, and expertise make it unable to function as a successful part of a sled team. Nevertheless, the husky puppy may be said to be thriving because it is showing the right signs of becoming what it is supposed to be. The more mature it becomes, the more it will show signs not just of becoming but of actually being an awesome sled dog. Humans, too, can show early signs of being the people we were created to be and also evince that we are moving toward our full purpose. Thus, thriving has both dimensions of right-now and not-yet-but-becoming.

Becoming what one is meant to be, or thriving, for any species is worked out in relation to that species' nature, including its telos. The husky thrives in the fullness of its husky-ness, what it means to be fully husky. If thriving involves approaching the fullness of one's nature, we need an account of what constitutes human nature.

HUMAN NATURE

Generally, human nature refers to the distinguishing characteristics of humans, but defining *nature* is fraught with difficulty, particularly when it is contrasted with *nurture* or *culture*. Many people think of *nature* as a kind of genetic or biological code that determines some aspects of who we are, whereas *nurture*, including enculturation and learning, determines other aspects. However, the nature-nurture distinction crumbles upon scientific scrutiny. Part of human nature is to nurture and be nurtured, and part of it is to be social and cultural; that is, humans are born into and develop in interaction with others who have their own particular diet, language, customs, and values. Nevertheless, it still seems that some common features of what it means to be human transcend individual and group differences in culture—there is still purchase to the concept of a "nature" that binds humanity together. To capture the idea that typical humans will develop in predictable ways, I adopt a particular sense of nature derived from philosopher Robert McCauley's discussion of various types of natural features.[2]

McCauley notes that essentially all humans of all cultures become very good at certain ways of thinking and acting. For instance, as a matter of ordinary development, humans become fluent in at least one language. Typically, humans do not have to consciously consider how to construct sentences, conjugate verbs, or select words in their native language—we simply speak. In an important sense, speaking a native language is natural. Native language use is characterized by fluency, automaticity, and ease. The same could be said for numerous other abilities such as seeing, hearing, walking, grasping objects, and reading emotional states from people's faces. Under normal conditions, these activities do not require conscious attention; we humans can just do them. McCauley observes, too, that of these abilities that become natural in the sense of fluency, automaticity, and ease, a large share of them seem to develop without the use of specific artifacts (i.e., human-made things) and without explicit instruction. Unlike riding a bicycle, walking does not require any artifacts and we do not explicitly instruct our children in how to do it. Unlike writing, fluency in one's native language does not require any particular artifacts or explicit instruction. McCauley also notes that these natural ways of thinking and acting are the sort of thing that typically develop very early in life, and if they do not develop, parents are likely to seek medical attention for their child. If a child is five years old and not speaking fluently, we grow concerned. If a child is five years old and has not become an accomplished writer, even after hundreds of hours of lessons and practice, sane parents will not rush to their family physician.

---

[2]McCauley, 2011. I add to this sense of "nature" in later chapters, but I find McCauley's approach helpful for getting started.

To summarize, McCauley suggests that some capacities and abilities may be maturationally natural,[3] as identified by the following features:

- They are characterized by fluency, automaticity, and ease.
- They develop early in life, typically during the preschool years.
- They do not require direct, explicit instruction to acquire.
- They do not require particular artifacts to acquire.
- Their absence creates concern about an underlying medical problem.

These maturationally natural capacities will be present in nearly all members of any cultural group and common across essentially all cultural groups. For the sake of our purposes here, I will simply refer to maturationally natural features of humans as "natural" and consider them part of human nature, even if they may not be necessary for or exhaust human nature. My task here is not to exhaustively demarcate what human nature is or essentialize what it means to be human. Rather, I am using the available science to serve as a pointer concerning the characteristic ways in which human nature is typically manifested in this life. By no means does it follow that someone with a developmental disorder or who has suffered injury or illness is not human or deserving of basic human rights and respect. Nevertheless, our appropriate concern with the full range of manifest humanity should not lead us to ignore the fact that there are typical ways humanity is expressed that, from an evolutionary perspective, have come about in part because of their service to our ancestors in solving certain types of problems.

Human nature, then, includes the psychological features of humans—ways of thinking and behaving—that develop in a maturationally natural way, along with the physical features of humans that are likewise the typical result of ordinary developmental conditions. Such an approach to identifying human nature avoids a false dichotomy with "nurture." All human nature is nurtured. Likewise, such an approach allows for

[3] Developmental psychologist Michael Tomasello employs a similar concept he dubs "maturationally structured learning" (2019, p. 34).

cultural particulars to fill in the contours of natural capacities, such as language learning. Proper language acquisition requires exposure to a language, and that particular language is determined by a cultural context, but that does not mean language use is not natural for humans. As with all aspects of human nature, it has historically come about in interaction with a particular sort of environment or niche.

FITTING IN A PARTICULAR NICHE

Without theology, our analogies with the husky and the Christmas tree are problematic. These two organisms have been selectively bred, domesticated, and reared for particular purposes, so a big part (but not all) of their nature is a result of human intentions and purposes. They have human-given purpose(s) much like we have God-given purpose(s), but what if we leave that aside for the time being?

What happens if we assume that God has not purposefully created us or any other kind of animal? Is there a way to talk about having a purpose or telos from a strictly naturalistic, evolutionary perspective? Let us explore this way of talking about thriving to see what lessons can be learned, doing our best to leave theology aside but bearing in mind that I do not intend to leave "thriving" as a theologically bereft concept.

Consider a nondomesticated canine that has not been bred by humans for a purpose: the wolf. What does it mean for a wolf to thrive? One way to think of a thriving wolf is simply a wolf that solves the problem of propagating its genetic inheritance successfully, as wolves do. From a naturalistic, evolutionary perspective, species only have "natures" to live up to in a weak and vague sense. As boundaries between species are not sharp, "nature" refers only to the general way in which a particular genetic population (species) has typically developed physically and solved survival and reproduction (fitness) problems behaviorally. So, for a wolf living in cold northern latitudes, evidence of thriving probably includes a thick coat of fur, strong bones and teeth, a place in a wolf pack, the ability to attract a mate, speed, agility, good hunting skills, and so forth. In short, a thriving wolf is a wolf that actualizes its nature in such a way that it wolfishly fits adaptively within its ecological niche, the environment in which it must live. Thriving, then, is partly an expression of nature and partly the fit of that nature within a niche. A wolf's thriving could be hindered by some failure to embody its nature (e.g., because of injury or developmental pathology) or by finding itself in an ecological niche for which its nature is poorly suited (e.g., a suburban neighborhood). Indeed, a wolf's ability to manifest its nature depends in large part on its niche. Practically, the nature and the niche are not separable.

From this perspective, humans, too, could be said to be thriving insofar as their nature and niche adaptively fit together. If a human solves the fitness problems of propagating its genetic inheritance (that is, mating and ensuring its offspring do likewise, in a humanish way), that human is thriving, right? Is it that simple? I think not. Evolutionary psychology gives us reason to think the story is not so simple for humans and that a big part of the complexity for humans has to do with how we humans characteristically address the fit between our nature and our niche.

All species impact the ecological niche of their offspring, otherwise known as niche construction.

Niche construction. All species impact the ecological niche of their offspring, otherwise known as niche construction.[4] For instance, fir trees change the acidity of the soil and the amount of shade in which their offspring will attempt to grow. Firs drop needles that decompose into the soil and thereby change that soil. As a tree grows it changes the amount of light available on the forest floor for its own seeds. More dramatically, a pair of beavers can actually change the local geography in which their pups will grow, including changing the number of trees nearby (by felling them), the speed and depth of nearby water (by building dams), and the amount of food nearby (by eating the best stuff first).

[4]Laland, Odling-Smee, & Feldman, 2000.

Humans are the most extreme niche constructors, engaging in a degree of niche construction that is unrivaled. We make clothes, plant crops, create elaborate tools, dig wells, and build villages and cities. Consequently, even though humans may not have genetically changed dramatically in tens of thousands of years, we have remarkably changed the environments in which our children must attempt to develop, survive, reproduce, and rear offspring, even in just the past dozens of years.

For instance, ten thousand years ago, a young human may have needed to learn how to chip a knife out of stone and use it as a tool. Now many young humans need to learn how to use digital technologies. Today, a man and woman who grew up having to hunt, fish, gather, and farm in order to survive can have children who have to manage spreadsheets and growth charts, navigate the online marketplace, and figure out tax forms in order to provide for basic needs. The skills and attributes required of humans today can be vastly different from what was required just decades ago. That is serious niche construction!

In contrast, ten thousand years ago, the typical timber wolf had the same problems to solve as it does today (barring human intrusion). The location in which the young wolf has to thrive might be slightly different from that of its parents, and the particular distribution of food sources, composition of the pack, and local microclimate might also be a smidgen different, but being wolfish, as wolves have been historically, can still be a good bet for a wolf to thrive.

Human solutions to survival problems, on the other hand, may leave lasting marks. Consider another contrast: a human versus a grizzly bear fishing for salmon. For a grizzly bear wishing to catch a salmon, good vision, quick reflexes, and sharp claws and teeth will serve her well. Over her lifespan, experience may improve her skills and perhaps these skills will benefit her cubs because they will learn from her. Her impact on her cubs' niche will be minimal and the bearish solution to catching fish does not radically change. Grizzly nature suits the grizzly niche.

Now consider a human who grows up catching salmon by hand.[5] Visual acuity, attention, patience, and quick reflexes are all valuable parts of human nature that could be put to this task. Initially, the human and grizzly solutions are similar, even if grizzlies have a few natural advantages (such as long sharp claws). Imagine, however, that one of these fishing-by-hand humans learns to use a fish trap or net to catch salmon. Suddenly the skills and attributes needed to catch salmon are very different. The human solution and the grizzly solution no longer look similar. The children of this human angler would never need to know how to fish with their hands, as few of us do today. Instead, knowing how to make and bait a fishing trap or construct and use a net would become a critical part of fishing. Both the physical and mental attributes of a successful angler would be different. If children are living in a fishing village and their thriving depends on fishing, what it takes to thrive has changed markedly in a short period of time, and the ordinary, natural toolkit with which they grew up may no longer be sufficient. Some human-made tools and cultural expertise are now required, so the solution to solve the problem of fish catching has, in turn, created a problem of toolmaking and teaching others how to make and use those tools. This is an example of niche construction that is common to the human species. The human niche includes the tools, techniques, and technologies we develop and make available to our offspring and not merely the changed physical environment, as with the beaver's dam.

It is easy to see that, over time, solutions humans have invented to address survival, reproduction, and child-rearing challenges may incrementally change typical human niches to the point that huge gaps exist between human niche and human nature. New technologies breed the need for new technologies. Humans' propensity to optimize their experience creates new challenges. Yesterday's success or innovation becomes tomorrow's new hurdle. "Progress" changes the demands on future generations that may be less natural for humans than previous solutions

[5]Yes, humans can and do catch fish by hand. Trout and salmon "tickling" and catfish "noodling" are examples. My grandfather grew up catching fish in this manner.

to these same problems. The gap between nature and niche is widened. I contend that these gaps augment and sometimes threaten human thriving: when humans expand the nature-niche gap, thriving is compromised; when we close the gap, thriving is facilitated.

Niche construction and thriving. Human thriving takes on a different character when we think of it in terms of fitting into an environment or niche. Instead of thinking about what it means to be fully human in the abstract, we may consider how our humanness meets, or fails to meet, the demands of a particular niche. In other words, what does it mean to thrive in actual and particular environments? Theologians Miroslav Volf and Matthew Croasmun (2019) argue that a component of flourishing is whether one is at peace or shalom with one's environment. One dimension of flourishing—or thriving—concerns a life going well within one's environment. So are there specific ways of being human that change our environments and make it more or less difficult to thrive? From an evolutionary-psychological perspective, are there specific features of human nature that have encouraged us to "niche construct" away from our nature, and can we close that gap and thrive? In subsequent chapters I discuss three clusters of capacities present in humans to an extent atypical of animals and, thus, a good place to look for specifically human means of thriving. These clusters are self-control, (hyper)sociality, and expertise acquisition. By self-control, I mean the peculiar human ability to act when we do not want to and refrain from acting when we do want to act; we can inhibit and channel our actions to satisfy desires and even change our desires. We humans can weigh future actions, chart new courses, and deliberately shape ourselves to an extent unseen in other species. Human sociality, too, is more elaborate than in other species. We live in large groups of nonkin, enjoying asymmetrical, personal relationships with more than a hundred others

When humans expand the nature-niche gap, thriving is compromised; when we close the gap, thriving is facilitated.

and less-personal relationships with sometimes thousands more.[6] We are enormously dependent on each other for getting information relevant to our fitness, including social information. Because of our sociality and our self-control abilities, we can also develop expertise by sharing in tasks, specializing what we do, and spending time learning from each other. We consume, process, and use enormous amounts of information, typically through language, and can thereby develop expertise not everyone shares. These three overlapping domains of competencies—sociality, expertise acquisition, and self-control—enable humans to both create and bridge the gap between their nature and their niche, a point I elaborate in chapter three.[7]

If the elements that both create and close the human nature-niche gap are the deployments of the same three groups of capacities, then merely exercising the capacities themselves—employing our human nature—is not thriving. Indeed, this reality suggests that what is critical is *how* these distinctively human capacities are used, not simply *that* they are used. As noted above, use them in a way that expands the nature-niche gap and thriving is compromised; use them in a way to close the gap and thriving may be facilitated, but there is likely more to the story than that. From a strictly evolutionary perspective, we have very limited resources for addressing the normative question of how they should be used. This question is one of aim, direction, and purpose.

---

[6]In relation to the fundamentally social character of humans, Pam King has written extensively on the telos of the reciprocating self. See Balswick, King, and Reimer (2016); King (2016, 2020, in press); King and Defoy (2020); and King and Mangan (in press).

[7]After arriving at this three-part method of organizing human distinctiveness (see King et al., 2017) and shaping this book around it, I discovered that Michael Tomasello (2019) stresses three similar groupings of traits to discuss what makes humans human. He goes into much greater detail concerning the particular forms of information processing, the way they develop, evidence for their development in humans, and differences with nonhumans. In the place of human sociality, Tomasello offers that human "social uniqueness is structured by the maturation of children's capacities for shared intentionality" (p. 8), which includes the ability to think and act in terms of a shared "we." Similar to what I present as expertise acquisition, Tomasello stresses "culturally learning from adult pedagogy as such and developing new skills through coordinative interactions with peers" (p. 8). He writes, "The third set of processes are humans' various forms of executive self-regulation" (p. 8), which closely maps what I am calling self-control.

MINDING THE GAP

One aim of this book is to step back and see if we can make some broader observations about what it means for humans generally to thrive—to detect general principles that might help (or hinder) people in their thriving, whether they are born to be mothers or merchants, brothers or farmers. To that end I observe that part of any species' thriving will be influenced by its nature and its fit with its characteristic niche or environment: a good fit between nature and niche facilitates thriving, and successfully bridging the gap between nature and niche moves us toward thriving.

What, then, is human nature? For much of this book I discuss some of the more prominent features of human nature when viewed in comparison with other species and from the perspective of evolutionary psychology. I focus in particular on three clusters of capacities: sociality, expertise acquisition, and self-control. The exercise of these three groups of features is a big part of what it means to be human instead of a trout, dog, or chimpanzee. We use these features to meet the challenges of our ecological niche. It is also the use of these features that may transform our niche, creating new obstacles for thriving.

This analysis of human thriving raises some challenges. By their nature, humans rapidly and dramatically adapt to problems in their environments and, as a result, change the environmental demands or niche of their offspring. Consequently, human nature in action constantly attempts to span nature-niche gaps only to create new ones. The result is a seemingly perpetual struggle to thrive. But is the situation quite so dire? In the next chapter, I further unpack just why it is that humans seem to struggle to thrive in a way many other species do not.

3

WHY AREN'T WE THRIVING?

Nature-Niche Gap

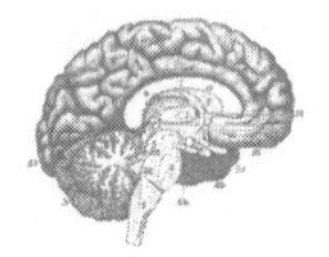

In the previous chapter I introduced the idea that human nature is usefully thought of in terms of the characteristics and capacities that arise as a normal part of human development in any cultural context. Take a baby human, for example: add the usual stuff from its environment and you get a whole lot of human nature and a whole lot of thriving, right? Actually, thriving is not that easy—it is not as simple as naturally growing up as a human. All of us do that. I suggest that thriving from the perspective of evolutionary psychology in part involves a good match between our nature and our environmental conditions, or niche.[1] Developing a good fit is the tricky part.

In this chapter, I further develop the idea that thriving, from a naturalistic, evolutionary perspective, may be understood as a matter of fine-tuning our natural capacities to fit our particular niche. Below I say a bit more about what human nature is, why it is that way, and, consequently, why we seem to confront gaps between our nature and our niche.

[1]I do not mean to imply that a good match between nature and niche is all there is to thriving or that one simply cannot thrive unless there is a strong fit. The case is analogous to having healthy teeth: good genes, a good diet, brushing, and flossing may all be involved in healthy teeth, but one may have healthy teeth even without one or more of these factors.

WHY IS OUR NATURE THE WAY IT IS?

Our focus here is on the psychological capacities that characterize human nature as opposed to the physical characteristics, such as having an opposable thumb or being a relatively hairless mammal. Nevertheless, in order to understand why we humans have the psychology we do, it is helpful to comment on why we possess particularly human capacities or features more generally.

Like many or most Americans, I think one answer to the question "Why are we humans the way we are?" is "Because God intended us to be this way for various reasons." And like most scientists, I also think the dynamics of evolution by natural selection are a big part of the answer. That is, God created a cosmos with living things that gradually change over time to suit the demands of their environments. For now, I leave aside the particular way or ways in which God may have used evolution to bring about humans. Fair-minded, devoted Christians who are also committed scientists will disagree on the particulars. Nevertheless, whether one views human origins as free from divine tinkering, as a story of supernatural intervention in the natural order, or as something in between, most can agree that humans possess human features in part because those features were good for survival and reproduction in human environments of the past.

As with other animals, we humans are the way we are because of demands on our ancestors to survive, reproduce, and rear their children effectively. For linguistic economy, biologists refer to this ability of a species to pass on its genes in an enduring way as "fitness." Those capacities that contributed to our ancestors' fitness were more likely to persist in subsequent generations—including ours—than those that did not. Though the term *fitness* may conjure up images of muscular gym rats, use of the word primarily involves behaviors and only secondarily involves our physical features. For instance, good fitness in animals requires the ability to procure food, but animals get their food in numerous ways using very different bodies. Some animals, such as barnacles, filter their food out of water using feathery filtering appendages, and others, such as jaguars, hunt and kill with sharp claws and powerful jaws. Getting food is the aim, not growing particular body parts. Similarly, when it comes to other aspects of fitness, such as self-defense or reproduction, the particular body parts appear to be negotiable, but whatever bodily features an animal has, it must engage in successful self-defense, feeding, and reproducing. Because of the type of animal humans are, we must also engage in childcare. Making babies that do not survive to make more babies is not a sign of fitness. Fitness-promoting behavior is the goal.

If performing the right behaviors drives fitness, then whatever physical features facilitate those behaviors will become more common in the population (all else being equal). It is easy to forget that our brains are part of our physical makeup that facilitates behaviors. Brains are part of bodies. Furthermore, insofar as our brains and nervous systems enable thoughts and thoughts guide behaviors,[2] then the kinds of bodies that have the right kinds of thoughts that drive the right kinds of behaviors that solve fitness problems will become most common in our population over time. Thus, the demands of our niche encourage fitness-enhancing behaviors, and those behaviors are shaped by the kinds of bodies we have and the kinds of minds we have. So bodies (including brains) and minds (supported by those brains) are subject to selection pressure indirectly through the behaviors they make more or less likely.

The idea that modern human minds have been shaped by ancestral fitness demands opens up some fresh perspectives on the character of human minds and human psychology and gives rise to the field of evolutionary psychology.[3] Evolutionary psychology is the study of human thought and behavior using evolutionary principles. It does not start with the assumption that human minds are like blank slates that are simply written on by experience in a single person's lifetime, empty vessels waiting to be filled by "enculturation" or "one-size-fits-all" general

[2]Psychologists have identified many situations in which the opposite happens—behaviors drive thoughts—but for simplicity, I leave those aside for now.

[3]For an accessible introduction, see Dunbar, Barrett, and Lycett (2007).

problem-solving computers. Rather, evolutionary psychology begins with the assumption that human minds likely share similarities to other animals' minds and other biological systems we see in the animal kingdom, systems and minds that have been tuned to solve common problems of their ancestors' past. In particular, those animals most like us in relevant ways—social mammals and especially primates—are valuable sources for drawing comparisons. What, then, do the minds of these comparable animals look like?

An inspection of the animal kingdom's physical and behavioral features reveals that animals tend to have specialized traits to solve particular problems that have been historically common in their niches. The long, thin, sticky tongue of the giant anteater is a wonderful tool for an animal that feeds on ants and termites living in nests in the ground—its tongue is a specialized tool for its niche. Volcano barnacles, small filter-feeding animals that colonize rocks on the seashore, close up in hard shells when the tide is out. Their hard shells fit their niche well because they protect the barnacles from dehydration, being crushed in the surf, or being picked apart by sea birds. Niche-appropriate specialized traits are most obvious in physical features, but they extend beyond body shapes and types.

Evolutionary psychology is the study of human thought and behavior using evolutionary principles.

Behaviorally, animals are characterized by various instincts and learning programs that suit particular survival strategies in their niches. Rabbits, for instance, are notorious for frequent mating. Their abundant reproduction makes sense for an animal that is a relatively easy mark for raptors, weasels, felines, and canines. Male rabbits, then, have a fairly enthusiastic mating instinct that targets moving objects about the size of an adult rabbit, and female rabbits are constantly fertile. Fowl, such as ducks and chickens, are famous for their imprinting instinct: the vulnerable chicks need to stay close to a protective mother, so they instinctively identify "mother" as a moving object of approximately the right size and follow it around until they mature enough to part with her. Thus, in natural conditions in which the chick has hatched from an egg underneath its mother, the chick will usually imprint on the right mother. What's more, group animals such as sheep, deer, and bison have herding instincts—where most of their group goes, they go; if some start running, all start running. Why? Living in groups reduces risk from predators. Generally, group living improves fitness, but keeping a group together requires behaviors, like herding, that keep individuals physically together. Cataloging natural animal behavioral routines could go on and on.

The point of these illustrations is to emphasize that the behaviors of any given animal are a bundle of behavioral routines (instincts) that are triggered by specific objects or events in the environment (stimuli). The general solution for fitness problems is not an all-purpose learning device that simply absorbs information from the environment or uses trial and error experimentation until it arrives at the best way to act. Rather, animals have natural dispositions to solve different problems in different, distinct ways. If we acknowledge the fact that humans are animals and likely descended from animals (including other human animals) that solved fitness problems through bundles of instincts, then we should expect humans also to have numerous instinct-like ways of solving problems.

As I explain in chapter five, the human capacity to flexibly learn specialized information is one of our species' distinctive features. To be sure, humans have some exceptional social learning abilities that are importantly informed by personal and cultural particulars. Nevertheless, the human mind is comparable to a toolkit with a host of specialized subsystems that have developed to solve specific sorts of problems that are generally important in the sort of environments in which we have traditionally lived, including social learning problems.[4] For instance, as

---

[4]Even if the best solution to a problem is learning from others of one's kind, knowing when to do so and having the tools to do so effectively is a specialized learning system not shared equally by all animals (see Laland, 2017). That is, even if one is attracted to the idea that human minds are

social animals that live in large groups of other humans, we need to be able to distinguish one person from another. To solve this problem, we appear to have a specialized mental device for detecting and recognizing individual human faces.[5] This system also appears critical for reading emotional states, in discerning which direction a human is looking, and other information that may help explain and predict a human's actions. In short, faces are so important to our day-to-day interactions and even survival that infants get a "leg up" on this task by being born with great facility in processing human faces. From shortly after birth, babies selectively attend to human faces in their environment and even imitate facial expressions.[6] They do not, however, have the same natural ease with recognizing individual cars, camels, canaries, or carnations.

Another example of a specialized learning system is the ease with which humans form fear associations with snakes. Given how our natural psychology works, we humans naturally pay more attention to snakes than snails or sneakers. And if we see others react with fear to a snake, we are more likely to become afraid of them than most other things in our environment—even things that are much more dangerous to us. Why? It appears that we have a natural predisposition to fear snakes, likely because some snakes in our species' past were very dangerous to our ancestors' fitness. Even though the risk of a dangerous snakebite for urbanites in much of the world is very low, humans still carry the tendency to fear snakes.[7]

The complete inventory of specialized subsystems that develop naturally in humans is the subject of ongoing research and is, hence, contested. Strong candidates for mental tools that humans naturally possess include

---

characterized not so much by domain-specific "instincts" but by the ability to socially learn (and teach), the mental toolkit for the social learning we see in humans is itself a specialized adaptation. For some suggestions concerning what might be in this distinctively human toolkit, see Tomasello (2019). Notice that in referring to these mental tools, I use the term *natural* as opposed to *innate*—a highly contested term that often implies being present at birth or somehow inevitable and unchangeable. I am open to these mental tools developing in various ways. The primary point is that they do typically develop as part of being a human in typical human environments.

[5]Kanwisher, McDermott, & Chun, 1997.

[6]Meltzoff & Moore, 1983.

[7]Öhman & Mineka, 2001.

those for thinking about bounded physical objects (including chairs and rocks), minded beings, basic numbers and their relations, living things, and human faces. Mental tools for negotiating social relations and foundational moral intuitions concerning interaction with others may be natural even if the social and moral systems built on these foundations vary considerably across cultures. Humans, then, have natural psychological subsystems that provide species-typical ways of thinking and acting in a variety of domains.

Some of these natural ways of thinking and acting are shared with other species, but if we want to understand human thriving, our focus needs to be on human nature. Just because a feature is special to humans does not necessarily mean that that feature is especially important; nonetheless, a comparison of human nature with that of other animals is likely to be instructive.

What, then, are those features of human nature that seem to be distinctive of, if not unique to, humans and our most recent ancestors? Given what we said above about how bodies, minds, and behaviors all interact under selective pressure to maintain fitness, humans have a number of interacting characteristics—some physical, some behavioral or psychological, and some cultural—that distinguish them from other animals. Perhaps no one characteristic cleanly and uniquely marks out humans from any other species that has ever existed, but it is obvious that humans have a bundle of features that make them stand out in some important ways from other living things on earth today. Below I suggest some particularly interesting natural human capacities that are shared in quality or degree with few or no other animals and group them in three categories: physical, behavioral, and cultural.

PHYSICAL FEATURES

We humans have a number of physical characteristics that enable us to be the kind of creatures we are, often in surprising ways. For instance, we have those handy and strong opposable thumbs. Humans can easily touch their thumb to each of their fingers and strongly grasp objects with all five. We can precisely pinch and hold small objects between our thumbs and index fingers. Though other primates, too, have opposable thumbs (as do some other animals), humans have improvements in thumb strength and dexterity compared with other primates that would likely have enabled much greater facility with creating tools and using them, including accuracy in throwing rocks or pointed sticks. These strong opposable thumbs and corresponding dexterity also enable humans to interact with surrounding objects from early infancy in a way that facilitates learning about the world and stimulates brain development. Sometimes the opposable thumb is held up as the physical marker of primate distinctiveness (and as a sign of approval), but some other physical traits are more important in distinguishing humans from other primates, particularly because of their impact on human minds, behaviors, and cultures.

Bipedalism. The dexterity afforded by opposable thumbs would be of limited use if we could not also stand, walk, and even run on two feet with head erect and eyes forward. Very few animals are bipedal—two-footed. Most animals walk on four (or more!) legs, if they have legs at all. Being bipedal enables us to use our two spare limbs to carry, throw, and do other work while standing up in a position with a strong visual vantage point. As it appears that bipedalism preceded the appearance of the supersized brains modern humans have,[8] it could be that the increased visibility, mobility, and dexterity enabled by walking upright created new nutritional opportunities for our ancestors—such as improvements in hunting and fishing—that generated enough calories and the right fats and proteins to build big brains. And big brains with heavy skulls to protect them are easier to carry around if they are resting squarely on the shoulders and straight(ish) spine of a biped than hanging down on the front of a quadruped.[9]

[8]Falk et al., 2012.

[9]In addition to humans, many birds have very large brain mass–body mass ratios and they, too, are bipedal. Consider, for example, the very intelligent, large-brained crow and parrot families. Some rodents such as mice also have big brains compared to their bodies, but their brains (and heads) remain relatively light. It does not appear an accident that the animals with the largest

Bipedalism, however, comes at a cost. Walking upright changes the angle and shape of the pelvis, which has implications for childbearing. Babies, particularly those with large heads, are harder to birth with a biped pelvis, putting the life of the baby and mother at risk. Fortunately, thanks to an unusual skull with loose plates and gaps, human infants' large heads can slip through the birth canal and continue to grow considerably after birth. Humans are also born premature in comparison with many other placental mammals, which also facilitates birth. Shortly after birth, baby horses can walk and baby monkeys can climb onto and cling to their mothers, but baby humans just lie there wondering how to lift their giant heads.

Small gut. We humans have small guts compared with other primates. We devote a relatively small proportion of our body mass to our digestive systems. In general, large guts (including stomachs, intestines, and the rest) are associated with low-quality diets; larger guts indicate they're made for food that is hard to digest or relatively low in calories. Consider the cow with its famous four-chambered stomach: it has an enormous gut because it has to work very hard to extract nutrition from all that grass. That humans have such small guts suggests that we also have relatively high-quality diets. Another way to think of it is that we have outsourced much of the digestive process to food preparation and cooking. A small gut and consequent short time digesting food means humans are more mobile than other similar animals. As we all know, it is hard to get up and go with a belly full of food. Imagine if that big meal stayed with you twice as long or you had to eat twice as much. That is a lot more time on the couch! Interestingly, spending a lot of time and energy on bellies also reduces the ability to invest in brain development. Therefore, smaller guts may also be an important key to growing big brains.[10]

encephalization quotient (a measure of brain size accounting for the fact that bigger animals just have bigger brains to manage all of that body) are humans (who walk upright) followed by various species of dolphins, whose aquatic existence means they do not have the same support demands for their big heads (see Cairó, 2011).

[10]For more on the gut size–brain size relationship in primates, see Aiello and Wheeler's (1995) landmark paper.

Big brains. Even for large mammals, humans have large brains. Big animals (such as whales and elephants) have big brains to manage their big bodies. To factor out ordinary bigness and try to get a better grasp of the relative size of different animals' brains, scientists have developed the "encephalization quotient," which provides a measure of how much an animal's brain weight deviates from other comparable animals, accounting for typical body size–brain size relationships.[11] With an encephalization quotient of 1 representing a typical vertebrate brain, hippos have a score of 0.5, or half what is typical. Gray parrots have an average score of 1 and chimpanzees a bit less than 3, or roughly three times the average. On the really big-brained end are dolphins at 5 to 5.5 and humans around 6.5—six times the average and twice that of chimpanzees. Perhaps even more important is the sort of brain matter that is so large in modern humans. If we think of human brains as having grown from an ancestor with a brain shaped much more like a chimpanzee's, the growth is predominantly in the prefrontal cortex, roughly the part of the brain in the forehead and in front of the ears. If you compare the size of the prefrontal cortex adjusted for body size in humans versus other mammals, humans really stand out.

Whites of the eyes. A feature of humans that is easy to miss but may be very important is the coloration of our eyes. The whites of our eyes, or sclerae, are very large relative to the colored iris. This arrangement makes it relatively easy to see which way our gaze is directed. Humans use this information all the time to determine what others are looking at and even to communicate nonverbally with others. The prominent contrast between the sclera and the iris helps an observer determine the angle at which the eyes are directed. To determine where an animal without visible sclerae is looking, we have to consider the angle of its head. Humans can use both head and eye angles when trying to determine the visual attention of others, which is especially handy when jointly considering an object. If I had only head angle and body posture to go on, if you

---

[11]See Cairó (2011) and also Roth and Dicke (2005), who report slightly different EQs, such as chimpanzees coming in at only 2.5 with humans around 7.5.

and I were to sit across from each other with objects in front of us, I might not be able to tell whether you were looking at me or one of the objects. Large white sclerae solve this problem and help us know we are both looking at the same thing.

Many mammals, including dogs, horses, and rabbits, have pale sclerae, but they also have very large irises that make it difficult to determine eye gaze direction (let alone that rabbits and horses have eyes on the sides of their heads). Nonhuman primates do not have white sclerae. The appearance of relatively large whites of the eyes in human ancestry may have been very important for sociality, as it would have helped in knowing what others were thinking about.[12]

Long lifespan. Our final example of a remarkable human physical feature is more about human physiology than anatomy. Humans average very long lifespans, particularly when compared with other mammals. More remarkable still is the fact that human females live a long time after they are no longer able to bear children. The implication is that we humans have considerable time to invest in our offspring—including grandchildren and others in our social groups—when we are no longer investing in rearing our own babies. We also have a long time to learn and acquire skills and insights that we can pass on to others and share with our communities.

GENERAL BEHAVIORAL FEATURES

These various physical features contribute to the peculiar psychological and behavioral features that also set humans apart. These remarkable behavioral features include slow maturation, lengthy investment in caring for and teaching offspring, docility, exceptional sociality, and broad ability to acquire specialized expertise.

Slow maturation. Humans grow up very slowly. Primates, including humans, and whales are extreme among mammals in terms of the amount of time it takes to sexually mature. In many respects, humans are the most

[12]Kobayashi & Kohshima, 2001; Jessen & Grossmann, 2014.

extreme.[13] The time it takes for humans to reach full maturity in terms of physical, conceptual, and social abilities is longer than most mammals' entire lifespans. Human children are typically not even proficient walkers until their third year (hence the term *toddler*) and cannot be reasonably expected to feed and care for themselves in a minimal sense until several years later. In comparison, a two-year-old dog is already sexually mature and ready to rear a litter of her own.

For our purposes here, a particularly important aspect of this slow maturation is that humans have a considerable capacity to learn and adapt to their specific environments. As a general rule, once bodies and brains have reached maturity, learning slows. As humans are slow to mature, they have many years of concentrated learning that can happen, so they are far less dependent on biologically fixed behavioral programs or instincts than other animals.

Investment in offspring. If you are a typical insect or fish, the steps in parenting include the following: lay eggs. That's it. Other animals, including mammals, show considerably more care. Slower-reproducing animals with fewer offspring nurture their young by defending, feeding, cleaning, and even teaching them. Humans take parental care to an entirely new level. Not only do we provide for our children's basic physical needs for more than a decade (and sometimes two or three!), we spend enormous time and resources deliberately teaching them tremendous amounts of information about the world around them, including how to care for themselves and how to behave socially. At the age that most other mammals have already started families of their own, humans are just starting school.

Not only do parents invest in their children; adults invest in other people's children. Grandparents, siblings, aunts, uncles, and cousins traditionally all care for their grandchildren, younger siblings, nieces and nephews, and younger cousins or cousins' children. Some humans—including teachers, daycare workers, and pediatricians—even invest

---

[13]Jones, 2011.

their lives in caring for the children of unrelated people. As a species, we invest enormously in children.

Docility. Compared with other predatory animals, including primates, humans are docile creatures. We humans crowd together with strangers in classrooms, buses, sports arenas, and grocery stores all the time. We do not typically fight to establish dominance or commit acts of violence over mating opportunities or scarce resources in these contexts. If chimpanzees, for instance, were thrust into such situations, the consequence would be violent bedlam, and yet we "trousered apes" manage to avoid daily bloodshed.[14] Our ability to use technologies for brutal warfare and homicide overshadows the fact that under most conditions, humans are remarkably docile animals. Humans have selectively bred other animals for docility, but it may not be misleading to say humans were the first domesticated animals.

Exceptional sociality. In part because of our docility and many years of dependence on others, we humans are enormously social animals: we live in huge groups of nonrelatives and interact with large numbers of others as individuals; we develop and maintain personal relationships with around 150 people on average,[15] plus hundreds of additional acquaintances and less-intimate relationships; and we spend considerable time thinking about the thoughts and feelings of other humans quite automatically. Indeed, the degree to which we think about our own or others' thoughts may be uniquely human. A few other social animals, such as chimpanzees, dolphins, and elephants, may have some capacity to understand others as having thoughts, often termed "mindreading." However, essentially no scholar believes that even chimpanzees can think about someone else's thoughts about their own thoughts, as in: "Jim knows that Pam believes that he wants a dog," or "William and Sarah both knew that they thought Tyler's idea was a good one." This ability is

[14]Thanks to Pete Richerson for this example. It may be that the reason for our "docility" is not merely temperamental but an interesting interaction in temperament, an unusual degree of self-control, and our ability to internalize social norms that regulate our behaviors. See Boyd (2018, chaps. 2 and 7) for some discussion along these lines.

[15]Hill & Dunbar, 2003.

referred to as "meta-representation" or "higher-order mindreading" (because it involves trying to know what others are thinking),[16] and it is immensely important for the kind of sociality humans experience and the ability to develop culture, which I elaborate below and especially in chapter four.

Self-control. Humans' relatively big prefrontal cortex appears to play an important role in human social intelligence, including mindreading. For this reason, among others, it has been suggested that our huge brain developed in our ancestors precisely because of its role in helping us navigate social life, which leads to better fitness. This part of the brain is also known for playing a related role in self-control. When we try to exercise self-control, focus our attention, and avoid overreacting to a burst of anger or frustration (leading, in part, to the docility mentioned above), the part of our brain behind our foreheads is likely doing some extra work. This deliberate, conscious regulation of our own thoughts, feelings, and actions enables us not only to navigate social relationships and cooperate with each other but also to think through multiple steps and plan for the future instead of acting impulsively. So in addition to increasing social intelligence, these massive forebrains enable self-control, which in turn helps us to coolly and calmly acquire and use information.

Expertise acquisition. Being a species that invests in its young and is also immensely social, somewhat docile, self-controlling, and slow-maturing creates the conditions for expertise acquisition. Children have the time and capacity to learn a lot from their elders and gain special knowledge that not everyone knows. Consequently, humans have enormous unevenness about certain types of knowledge they possess. Some know how to build a house and others do not. Some are skilled at making clothes and others can barely dress themselves. Some can cook wonderful food and others will burn breakfast cereal. Unlike essentially all other animals, skills and knowledge related to basic survival needs are radically uneven in humans.

---

[16]Baron-Cohen, 1995.

In times past, this degree of specialized expertise was probably considerably lower: when it came to basic survival, most people of the same sex had a common base of knowledge and skill, with perhaps one or two tasks or domains requiring particular skill or expertise. Today, in complex agricultural, industrial, and postindustrial societies, the degree of expertise acquisition and role specialization is astounding, a point we return to at the end of this chapter when considering purpose in life in relation to thriving.

Moral thinking. The human ability to mindread and form thoughts about thoughts (meta-represent) enables us to reflect on our own or others' motivations for actions and decide whether actions are good or bad on that basis—not just on the basis of the consequences of those actions. For other animals, whether something is good or bad (if they can even think about good and bad) is likely related entirely to consequences, including reward and punishment. Animals will learn to engage in or avoid certain behaviors depending on whether or not they are followed by a positive or negative experience. Humans also do this, but we may be the only animals that consider whether actions are good or bad regardless of whether they lead to immediate reward or punishment. We consider hypothetical future actions and evaluate them as good or bad in relation to some abstracted, externalized "objective" standard. We do not merely consider whether you might approve or disapprove—a dog or chimpanzee can anticipate if it will be attacked by a higher-status individual if caught doing something the superordinate doesn't like—but we consider whether the action is objectively (in some sense) good or bad regardless of whether another person approves.[17]

Often part of our evaluation of an action is a consideration of the perceived motives or intentions behind the action. For instance, if someone gives you chocolate-covered hazelnuts, you will interpret the gesture as

[17]Tomasello sees this capacity as a product of what he claims to be the uniquely human ability of "collective intentionality" (2019, p. 77 and elsewhere), the ability to put together many multiple perspectives on an object, event, or behavior and thereby construct a collective viewpoint, easily understood as a relatively complete perspective from nowhere or an "objective" vantage point. Tomasello argues that the strongest evidence for this ability does not emerge until children are about three years old but that it never appears in other apes. We can use this collective intentionality to evaluate (and regulate) our own behaviors.

an act of kindness if you assume the giver meant it kindly even if you do not like hazelnuts or are deathly allergic to them. If you have reason to think the giver knows full well that you are deathly allergic to hazelnuts, then the same action would be regarded as an act of carelessness or even attempted murder. Knowledge and thoughts of the actor matter, and our very special mindreading capacities make these considerations possible.

Rationality. The ability to think about thoughts enables humans to exercise rationality, however imperfectly. That is, humans can reason and use logic to draw conclusions from beliefs and evidence. If Professor Plum, Mrs. White, and Mr. Green were playing billiards at the time of the murder, and Ms. Scarlet and Mrs. Peacock were both in the study, but the murder took place in the hall, then we can conclude Colonel Mustard committed the crime. Or, if we know that all birds have feathers and a whackadoodle is a type of bird, we can conclude that a whackadoodle has feathers. Upon our abilities to exercise rationality we humans have built mathematics, philosophy, theology, and the sciences. Rationality enables us to prune away flimsy and unjustified beliefs and discover useful truths.

CULTURAL FEATURES

Cultural features are those that are passed on through social learning and may vary considerably in their expression depending on from whom and where the features were learned. It is undeniable that humans have a degree of cultural expression like no other animal, including the domestication of other species, complex tool use, cumulative culture, and language use.

Domestication of other species. We humans may be the first domesticated species but we have played it forward. Humans have selectively bred many different species in order to enhance the traits useful to humans. These domesticated species include familiar animals such as dogs, cattle, and chickens, in addition to plants such as corn and wheat. As compellingly articulated by Jared Diamond in *Guns, Germs, and Steel,* this close relationship between humans and animals changed not only the plants and animals that were domesticated; it also changed humans.[18]

[18]Diamond, 2003.

Arguably, the domestication of plants and animals facilitated settled life, food surplus, specialization of labor, and technological innovation, in addition to changes in disease resistance and nutrition.[19] Whether for hunting, plowing, transportation, or any number of other uses, animals have been standard tools and companions for humans for thousands of years.

Tool use. One unmistakable feature of humans is that we are tool users. Lacking sharp claws or powerful fangs, we make blades and clubs. When our human ancestors migrated into colder climates, they fashioned suitable clothing and created dwellings. Perhaps most strikingly, humans start and control fires. Other animals fashion and make tools, including chimpanzees and crows, but the complexity and variety of human toolmaking leaves chimpanzees' termite fishing sticks looking rather unimpressive. How are humans capable of such technological mastery?

Physical attributes such as size, strength, upright posture, dexterous hands, and long lifespans surely contribute to the toolmaking prowess of humans: we can manipulate a wide range of materials because of our strength and dexterity, and a long lifespan allows for an accumulation of knowledge and skill that can be passed on to children and grandchildren. Passing on the "tricks of the trade," however, assumes a certain kind of mind—one that is eager to teach and can seek to learn from others.

Cumulative culture. Make no mistake, many other animals learn from each other's behaviors or engage in what might be called social learning.[20]

---

[19] I also wonder whether the sort of design- and purpose-based reasoning about the natural world that Boston University psychologist Deborah Kelemen and her colleagues have documented as being so natural to humans, particularly in childhood (Kelemen, 2004; Kelemen & Rossett, 2009), may in part be a product of domesticated plants and animals. Ponies and pea plants really do have the features they do for particular purposes, for which humans have selectively bred them. This tendency to see potential purposes in the natural world may have, in turn, led to more creative problem-solving and innovation using natural materials. Or perhaps some tendency toward design-based reasoning helped along the domestication of plants and animals, which reinforced and broadened that design-based reasoning. I wonder, too, what other psychological influences close living with animals has had on humans and how that might change with more distance from them because of urbanization.

[20] Kevin Laland (2017) describes some fascinating examples of how various animals, including some species of stickleback fish, modify their behaviors after observing others and uses these examples to explain how human social learning differs from that of any other animals that have been studied. In particular he stresses human cognitive and social mechanisms for high-fidelity transmission of information.

Other animals, particularly bonobos and chimpanzees, engage in high levels of social learning leading to behavioral practices that look very much like culture, including tool construction and use, customs of food gathering and preparation, and characteristic gestures. Humans, however, develop cumulative culture to a degree that is unique among existing animals. With other animals, we see an individual innovate—for example, a particular termite fishing stick design—and then that innovation becomes common in the group and a stabilizing part of the group's "culture." In humans, however, innovations both spread and change much faster. Humans build on each other's ideas. For instance, we humans readily combine tools to make a new tool or modify the existing tool to give it improved or new functions.

Consider a fairly simple, common tool, the knife. Humans have had cutting blades for tens of thousands of years and have also modified these blades to better suit particular functions: scraping, hacking, slicing, peeling, paring, boning, and so on. Humans have also taken the utility of long straight sticks and added the utility of a point or blade to them to create spears and arrows. We see nothing like this kind of rapid reimagining of tools in other existing species. The way in which basic food preparation techniques are combined or modified to create a dizzying array of cuisine, or the way in which kicking a ball between two uprights can evolve into a heavily regulated and globally popular sport, are other examples of cumulative culture.

Language. Though the particular differences between human language and dolphin or chimpanzee communication are debated, there is no doubt that the degree to which *Homo sapiens* is a deeply linguistic species is unrivaled. Many animals can signal to each other through vocalizations, other sounds, or even body language, as is famously the case with the honeybee's "waggle dance" that indicates the direction and distance of a good batch of pollen-rich flowers. Even very clever dogs can be taught hundreds of commands that lead to actions. Humans, however, do not just have a repertoire of signals that can be used to tell others to run or climb or submit; we can also communicate information with no obvious behavioral consequence, just because we want the other to know something.[21]

My uncle had a border collie that would herd any of a number of farm animals on command, including the sheep, chickens, and even ducks. "Missy, go get the ducks" sent this dog swimming out in the pond in an attempt to bring in the ducks. Missy was a brilliant and admirable dog, but she could not be told, "Missy, the sheep will have more lambs if you don't unnecessarily stress them," leading to a change Missy's mind or behavior. "Missy, the leghorns lay better than the other chickens" would be wasted breath. It seems only humans can expect each other to draw novel but predictable inferences from words and gestures. Humans alone inform each other in ways that do not directly correspond to immediate actions. We do not merely command each other or tell each other things that should immediately change our behaviors; we humans inform each other.[22]

The fact that humans can inform each other with language means we can also change each other's thinking with great flexibility and thereby accumulate knowledge. This knowledge accumulation and ability to communicate it in turn leads to cultural innovations. Thus, language is a critical catalyst for other cultural features.

Mindreading and cultural innovations. Because it is so important, it is tempting to say language sets humans apart as a distinctive animal and leads to all of their other noteworthy differences from other animals. Perhaps language is the lynchpin capacity of human uniqueness. Alternatively, it could be that our high-level mindreading abilities combined with

---

[21]In discussing one type of communication used by human infants, pantomime (iconic gestures in which someone tries to convey an idea by enacting a representation of it), Tomasello explains, "Presumably great apes do not understand iconic gestures because they do not understand communication marked ostensively as 'for you' (co-operatively). If an ape views someone hammering a nut, they know perfectly well what he is doing; but if they view him making a hammering motion in the absence of any stone or any nuts, they are perplexed" (2019, p. 107).

[22]Occasionally chimpanzees, bonobos, dogs, and parrots do things with language that are truly outstanding for their particular species, but in no case have there been well-documented linguistic feats in nonhumans that approach what we see in the typical four-year-old child. It is also worth noting that the linguistic geniuses of the animal world are (almost?) always heavily socialized and trained by human language users.

some basic linguistic capabilities created the kind of cultural animals humans are today. When two people know they share thoughts about the same object, they can use words to give each other information and not merely bark commands. When a parent and child are paying attention to the same object but have different perspectives on it, the parent can teach the child the name of that object and tell the child things about it. Mindreading takes basic language to a more powerful level, enabling humans to talk to each other about what they are going to do and how they will coordinate their actions to accomplish a task or improve on a plan. In this way, it may be that mindreading and language interact with each other to make humans especially effective at developing cumulative cultural innovations.

TO BUILD A HUMAN

The physical, behavioral, and cultural features of humans sketched above are not merely a grocery list of characteristics that humans and few other animals have. Rather, these features represent more of a recipe for how to make an animal that is completely unique in this world. What's more, these varied characteristics make one another possible. Physical features such as bipedalism and a small gut made possible big brains that include tremendous learning flexibility and social intelligence. Longevity and slow maturation give humans lots of time to learn about a particular environmental context, acquire cultural particulars, and amass knowledge and expertise that can be passed on to the next generation. Large whites of eyes and big social brains enable thought about others' thoughts, desires, and motivations, allowing for increasing cooperation, sharing of ideas, learning from others, and building on inventions and insight.[23] These special brains that can mindread and form thoughts about thoughts make a more complex form of language possible, which in turn facilitates

[23]Kevin Laland (2017) and Joseph Henrich (2015) provide well-developed and stimulating accounts of how the human cognitive abilities that underwrite acquiring information from each other have led to us becoming not merely social animals but cultural ones. Cultural environments, in turn, may have importantly shaped human evolution.

rapid cultural innovation. Thoughts about thoughts enable reflection on whether some ideas or behaviors are good or bad even before they are put into action. That is, this combination of features makes us humans into moral and cultural animals like no other.

As Pam and I discussed these various features of humans with colleagues in relation to how they might relate to broader psychological treatments of thriving, flourishing, or positive development, we found a helpful way of grouping many of the most notable capacities,[24] another way to summarize how humans uniquely deal with the challenges of life. Rather than only bite, claw, camouflage, burrow, run, climb, herd, or hide our way through life, we humans address the challenges of survival through unusually robust hypersociality, information gathering and use, and deliberate self-control. Or in Pam's shorthand, to survive and thrive humans relate, learn, and regulate. These three overlapping and mutually supporting clusters of features characterize our nature and enable us to tackle the demands of our specific environment or niche. Distinctively human thriving, then, involves in part the proper deployment of these features, a claim I will develop further in the subsequent three chapters.

To survive and thrive humans relate, learn, and regulate.

STONE AGE MINDS IN MODERN AND CHANGING CONTEXTS

Another lesson from this analysis is that the particular mental tools in our toolkits may have as much or more to do with ancestral problems than contemporary ones. As outspoken champions of evolutionary psychology, Leda Cosmides and John Tooby are fond of saying that "our modern skulls house a stone age mind." They explain:

[24]See King et al. (2017) for how we arrived at these three clusters.

> Natural selection, the process that designed our brain, takes a long time to design a circuit of any complexity. The time it takes to build circuits that are suited to a given environment is so slow it is hard to even imagine—it's like a stone being sculpted by wind-blown sand. Even relatively simple changes can take tens of thousands of years. The environment that humans—and, therefore, human minds—evolved in was very different from our modern environment. Our ancestors spent well over 99% of our species' evolutionary history living in hunter-gatherer societies.[25]

In short, our biological brains are not evolving or adapting at a rate that even comes close to keeping up with our changing niches. Thus, we have really old wiring! Their point is simple, but the implications wide-reaching: if our brains structure minds that are designed for problems of our human past, then they may misfit in some important ways with our present. Even if our minds are not as fixed by our biology as Cosmides and Tooby seem to suggest, in some fundamental ways our psychological nature changes much more slowly than our niche, resulting in a gap between what our minds are "designed" to do and the challenges that the ever-changing environment throws at them.

This nature-niche gap has not escaped the attention of nutritionists. An almost cliché example of our nature-niche gap is the strong human attraction to saturated fats, salts, and sugars. As mentioned in the introduction, as the story goes, in our ancestral past during which our food preferences evolved (a theoretical time known as the "environment of evolutionary adaptedness" or EEA), foods heavy in saturated fats, salt, and sugar would have been scarce but very good sources of nutrition. It

[25]Cosmides & Tooby, 1997, p. 12. In recent years some psychologists and other evolutionary anthropologists have taken issue with the idea of Stone Age minds but instead want to stress that the nature of human minds may have evolved very quickly due to cultural evolutionary factors (e.g., Laland, 2017). Note, however, that the Stone Age is really not that long ago. Many peoples of the world, including those on the Hawaiian Islands and many Native American and First Nations populations of North America, used only Stone Age technologies until the late eighteenth or even nineteenth century. In other parts of the world, the Stone Age didn't end until the twentieth century. Thus, unless one is prepared to argue that the minds of some humans are radically different from others due to cultural evolutionary dynamics, which I have not seen by any credible scholars, it is no exaggeration to say that our default biological equipment is Stone Age, and that is the equipment we need to use to survive and thrive in this modern context.

would have been adaptive to find such foods very attractive. Now that many of us live in a niche in which such foods are more abundant, we can easily overconsume them because we find them so attractive and thereby damage our health by inducing diabetes and heart disease. If not for other medical and nutritional advances, perhaps what was once fitness-enhancing would now be fitness-threatening because of a change in niche. Our nature has not (yet?) changed to account for our new niche. The rationale behind the currently popular "caveman" or "paleo" diet is that we should eat a diet that matches our ancestral conditions: simple foods, low in processing, emphasizing meats and vegetables. Even those solidly in the evolutionary camp may disagree over just which foods you should or should not avoid, but the rationale is fairly consistent: we have Stone Age bodies and minds living in a modern world. That is, a gap exists between our nature and our niche.[26]

Another example that highlights the nature-niche gap is antibiotics. We have developed a suite of antibiotic drugs that can be used to combat diseases and infections caused by various bacteria. On our own, many of us would have much graver and even deadly illnesses if not for these drugs. When our nature did not provide a satisfactory way of meeting the niche demand that various bacteria presented, we found a medical technology to bridge the gap. An unwelcome side effect, however, is that we have now changed selection pressure on these malignant bacteria. Strains that are more resistant to antibiotics are more likely to survive a course of treatment and spread. Thus, we have inadvertently encouraged strains of bacteria that are resistant to our medicines, requiring the search for new antibiotics. This dynamic is especially worrisome if our use of antibiotics pushes bacteria to become more resistant to even our bodies' natural immune defenses. We may end up with little or no natural

[26]The nature-niche gap between our minds and our niches may be greater than that between our guts and our niches because the properties of our guts are heavily influenced by the microorganisms that live in our digestive systems and enable us to process different sorts of foods. These microorganisms can change and evolve much more rapidly than humans can and the particular types of microorganisms that live within us can change with what foods we are consuming at a rate much faster than human nature can change.

resistance to many of these diseases and so we have to keep chasing new technological solutions. Hence, in some cases, new medical technology is required to solve the problem that our immune systems on their own are ill-prepared to meet.

These nature-niche gaps occur not only in humans but in domesticated animals, too. Let us return to rabbits mating, chickens imprinting, and deer herding.

In natural circumstances, rabbits live in colonies with few other animals of the same size within reach, so their mating instincts do the job nicely. The proper domain (i.e., the situations for which it evolved) for a male rabbit's amorous intentions is female rabbits, but the actual domain of objects that trigger mating behaviors can include male rabbits, small dogs (or large dogs' heads), and soccer balls. Even sleeping cats can become unwitting victims of rabbits' advances. By putting rabbits in a different ecological niche, such as single domestic rabbits in a back garden of a home, we can see a humorous and somewhat disturbing mismatch between their instinct and their niche.

Similarly, YouTube is full of charming videos of ducklings and chicks imprinting on cats, dogs, and humans instead of their own mothers. When fowl are hatched by humans or otherwise raised with domestic animals, their imprinting mechanism may select these other targets as "mother" and follow them around. The story may not end well when a typical cat or dog is involved, but even cats, terrifyingly adept predators, sometimes "adopt" these ducklings and chickens as their own brood, particularly if they have recently given birth to their own litter. The maternal rearing instincts of the cat sometimes override predatory tendencies and we find ducklings suckling alongside kittens on a contentedly purring mother cat.

Herding animals such as deer, horses, and sheep would naturally live in groups of their own kind, occasionally mixed with other similar herd animals. Their tendency to run the same direction when others run evolved under these conditions. Now that such animals live in close proximity to humans, their instinct can be triggered in peculiar and amusing ways. Deer will sometimes run alongside mountain bikers in a forest. Horses will run along with cross-country running teams. Whereas the proper domain for herding remains others of their herd, the actual domain of stimuli that triggers herding responses in deer may include humans running or mountain biking.

In each of these three cases, the animals in question have a natural way of thinking and behaving that gets redirected because of their proximity to humans. Humans have impacted these animals' niches in such a way as to create friction between nature and niche, and less charming examples exist as well. For instance, whitetail deer in the Great Smoky Mountains National Park have been free of human and most other predation for over eighty years, leading to weakened survival instincts. People killed and otherwise drove off most of the serious natural predators for deer (mountain lions and wolves). With hunting prohibited in the park, whitetail deer lack natural skittishness that would serve them well as prey animals. They had better stay in the park! In the neighboring Cherokee National Forest, hunting is permitted. Whitetails there do not show such casualness around humans.[27] So people change the niche of other animals, impacting their natural toolkits for surviving and thriving. Humans do this to themselves, too, abundantly.

We can change our environments or niches faster than our natures can change.

The three overlapping clusters of features that uniquely characterize human nature—being hypersocial creatures, massive information users who acquire expertise, and deliberate self-controllers—are at once the ways we solve problems in our niche and the drivers of a greater gap between our nature and our niche. We use these features to change our environments in order to solve problems, which in turn creates gaps between nature and niche. We can change our environments or niches faster than our natures can change.

[27]I thank Drew Crain for this example.

How might these nature-niche gaps impact human thriving? Consider finding purpose in life.

AN EXAMPLE: FINDING OUR PERSONAL PURPOSE

The concept of purpose has pervaded cultural and academic realms. For example, Rick Warren's bestseller, *A Purpose Driven Life,* resonated with so many readers because it identified and wrestled with a central problem for many contemporary people living in the developed world.[28] People commonly wonder, "Why am I here?" and "What is my purpose in life?" Not surprisingly, a wealth of research on what it means for teens to thrive has documented that a sense of purpose that helps define a teen's self-concept or identity is critical.[29] Teens wrestle with whether or not they should go to college, what they should major in once they get there, what career they want to pursue in adulthood, where they want to live, and what their goals in life should be. Then they have to consider how those goals motivate decisions about what classes to take, what clubs to join, who to befriend, which sports to invest in, and so on.

A sense of one's purpose looks like a key ingredient in a thriving life, but is it always? Has it always been? Consider a very different ecological niche for a human. Imagine living ten thousand years ago in a village of approximately 150 people for whom daily activities typically involve fishing, hunting and gathering food, food preparation, eating, clothing making and repair, homebuilding and repair, toolmaking, handicraft and visual art, dancing, singing, storytelling, and engaging in religious rituals and ceremonies. Before we unduly glamorize this life, we should also note that shoring up defenses against storms, predators, and enemies is likely a part of this life. A violent death could be around the corner.

Given this very different ecological niche, let us reconsider one of those big questions about purpose in life that contemporary children and teens ponder: What should I be when I grow up? Only some cultural conditions present a range of options concerning what your vocation

---

[28]Warren, 2002.

[29]See, for example, Damon (2008) and King and Argue (2020).

might be, and only some cultural conditions equate employment with what you are. To hunter-gatherers, the question "What do you want to be when you grow up?" would probably be absurd. Typically, boys and girls will be men and women, fathers and mothers, members of a pre-ordained people group, with a few skills shared by others of their sex and clan. What they will be is a member of a community that needs them to contribute for everyone's sake.

Deciding and charting one's own purpose is an important accomplishment in many contemporary societies, but this is largely because these societies have created a degree of career specialization and individualism that is peculiar in the history of our species. That is, these contemporary niches demand discovery of purpose for thriving, but our nature appears to be ill-equipped to find a purpose that defines identity in some conscious, deliberate, idiosyncratic way. This results in millions of people struggling to find purpose in their lives, whereas, in past generations, that purpose was almost entirely supplied by their parents, clan, or other features of the cultural environment.[30] Ancient Hebrews, for instance, knew they would adopt the roles and vocations specified by their family and clan and that an umbrella purpose for every person was "to do justice, and to love kindness, and to walk humbly with [their] God" (Micah 6:8). It may be that someone would have an additional idiosyncratic purpose, but this would be given to that person through prophecy or divine encounter and not sought out for oneself.

The competing values and vocational choices within the cosmopolitan and rapidly changing contemporary world open up genuine questions about one's individual purpose. Consequently, we must marshal our distinctively human abilities to gather cultural information and expertise, manage and navigate an enormous number of social connections, and exercise self-control in order to discover a satisfactory purpose, one that

[30]Just because different vocations were not as apparent in ancient or former cultures, these people might have felt purposeful about their work and daily lives as means to using their skills to contribute to their family or people group.

builds on one's strengths and passions while contributing in a culturally meaningful way to the good of others. That is, because human methods of solving problems have created such a diverse and complex niche, those same distinctively human psychological capacities have to be leveraged to discover a noble personal purpose in order for us to thrive. More specifically, our deployment of sociality, ability to learn, and self-control have created such a complex niche that we may have too many relationships, opportunities to learn, and potential goals to pursue. And so we need to use those same skills to find and select a meaningful purpose to pursue in order to effectively focus our relationships, learning, and self-control in order to thrive. I detail these three clusters of capacities in the next three chapters.

4

SOCIAL GAPS

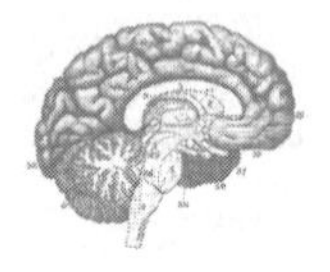

Evolution is often misunderstood as premised on a kind of rugged individualism, that it is "survival of the fittest" with every animal in competition with every other. Do something nice for someone else—give them food or protect them from an attacker—and you reduce your own fitness. Such a view of evolution, however, is mistaken. In many different animals we see behaviors that look a bit like altruism in humans. This apparent altruism ranges from mothers risking their lives for their young, such as when mother rabbits attack snakes (yes, bunnies go after snakes), to wolf packs sharing a kill, to honeybees committing suicide attacks on threats to the hive.

Many animals are social and improve each other's fitness at the expense of their own individual livelihood. When worker honeybees sting an intruder, dying as a consequence, they are laying down their lives for the rest of the hive. Also, worker bees give up the chance to reproduce for the sake of the queen, who is the only one who mates (with drone clones). The individual fitness of bees appears to be forfeited for others. Why? A beehive is populated by worker bees that have the same mother (the queen), and many also share the same father (drone) who shares that mother (the queen). Honeybees in a hive share a high degree of genetic relatedness with each other, and so, even if they die without reproducing, if their behavior benefits the hive, their genes carry on. So the "altruism" of the honeybee does not necessarily mean loss of fitness. If fitness means passing one's genes on to the next generation, then sharing food or even laying down one's life for the sake of others can increase fitness rather than compromise it, provided those others share enough of one's genes. This type of fitness is often termed "inclusive fitness" to capture the idea that it is not a given individual's fitness that natural selection will favor but the fitness of an individual along with its close relatives.

Taking care of those related to you, or kin altruism, appears to be governed in many cases by what is called "Hamilton's rule": the idea that altruism can evolve in a population if the cost of the altruism is less than the benefit to the recipient multiplied times the degree of relatedness.[1] That is, the more closely an organism's genes are to another's, the more that organism will sacrifice for that other. Parents will do a lot for their children because a child carries on average half of a parent's genes. Siblings will tend to look out for each other because they average half of the same genes in common. Aunts, uncles, nieces, nephews, and cousins receive a bit less self-sacrifice because they share a smaller degree of relatedness, and more distant relatives or nonrelatives would be even less likely to receive serious sacrifices as a general principle.

These kin-based selection principles can account for a lot of what looks like "sociality" in nonhumans. Nonhuman animal groups often have a very high degree of relatedness among their members. What about humans? How do we compare?

Humans are immensely social animals. If you go into a pet store that carries aquarium fish, you may find that certain ones are labeled as "social" or "community" fish. Such fish tend to swim in schools and may show signs of anxiety if not in a school of their own kind. We sometimes consider other animals as "social," including pigeons, deer, cattle, and sheep. These animals also tend to live in large groups rather than strike out on their own. Nevertheless, our use of the label "social" for such animals

[1]Hamilton, 1964.

should not obscure the fact that human sociality has a very different quality from that of most, if not all, other animals.

We humans do not merely prefer to live in communities with unrelated humans as herd or pack animals do, but we also form many unique relationships with others in our community. We treat specific humans as distinct individuals, we spend more time with some compared with others, we anticipate their particular desires and foibles, we remember past interactions with them, and we form emotional attachments to them. An adult gazelle in a herd will essentially treat all of the other gazelles the same way, and if a cheetah picks one off, no tears are shed. Not so with humans. And on top of treating many others as distinct individuals, we also can form a sense of "we" that motivates common goals even with distinction of roles within those goals. It is not clear any other animal does this.[2] Humans are hypersocial animals.

SOCIAL BRAIN HYPOTHESIS

Human sociality is so striking that a leading account of why our ancestors evolved such large brains is called the "social brain hypothesis."[3] It turns out that the size of the outer layer of our brains, our neocortex, including the important prefrontal cortex, is a strong predictor of group size in primates.[4] Primates with a larger neocortex either in absolute terms or as a proportion of total brain size form bigger communities. Furthermore, this evolutionarily "new" part of the brain is remarkably large in humans. Humans have big neocortices and, thus, big groups. This fact, among others, suggests that humans may have unusual brains because these brains improved our ability to live cooperatively as social animals, exchanging information with high fidelity and sharing resources. This bundle of capacities led to improved fitness.

Information about others. Among other things, our big neocortex, especially the prefrontal area, enables much greater degrees of social

---

[2]Tomasello, 2019.
[3]Dunbar, 2009.
[4]Dunbar, 1993; Barrett, Dunbar & Lycett, 2002.

intelligence than that found in other species. We use this brainpower to identify individuals, learn information about them and from them, store it, and retrieve it. Consider a good friend of yours. How would you describe this person? You could list distinctive physical traits, such as being tall or short, dark or fair; features of disposition, such as being outgoing, energetic, emotional, or conscientious; likes and dislikes; passions and interests; skills; personal history; and who friends and family members are. You may even periodically take note of this friend's current likes and dislikes and attitudes toward other individuals, such as being angry toward a sibling or smitten with a particular someone. You have enormous amounts of information stored that picks out one and only one person from the billions on earth. And, if you wanted to do so, you could probably describe a hundred other people you know to a similar level of detail. That is some social brainpower!

Even more importantly, humans can take this information and use it to predict and explain new behaviors and to make decisions based on these predictions and explanations. Suppose I need a lift to the airport. Who do I call? My decision-making capacity will draw on what I know about my friends: "Gary knows all of the best routes and how to avoid traffic, but he drives a little fast and that scares me. Rebecca is eager to help and very patient in heavy traffic but does not like mornings and would probably scold me for not arranging a ride sooner. I know! I'll phone Greg. He is dependable and cheerful at any time of day." We use social intelligence, some of it automatic and not accessible to conscious inspection, to decide who to ask about certain types of information and who to copy or learn from. Such decision-making ability based on our social intelligence is so automatic that we rarely even notice it, but imagine if we humans did not have this information or could not use it. As a species, the human would be an entirely different sort of animal.

Mindreading. We humans combine our knowledge of other individuals with situational knowledge. For instance, suppose a friend comes to your home to visit and you have his favorite snack in a cupboard. Unless you know that your friend has seen the snack in that cupboard before, you know you have to direct your friend to the proper location of the snack and not just say, "I have your favorite snack. Help yourself." That is, adult humans understand that beliefs about the location of a desired object are formed through testimony from others or through perception via the senses. Lacking these sources of information, it is likely they would not know where an object is. Your friend would begin searching arbitrarily if not told the proper location. Interestingly, a typical three-year-old would not be able to say why someone looks for a favorite snack in the wrong location instead of where it is really located. At that age, a child's understanding of how other people's minds work is not mature enough to comment on such a mundane task. It may be that even adults of other species cannot solve this sort of task. Indeed, even describing the situation may have struck you as peculiar, in part because to the typical adult such details in our social interactions are handled so automatically that they do not even merit attention.

This ability to both track the mental states of others, including their beliefs and desires, and use this information to predict and explain their actions has been termed "mindreading," "theory of mind," and "mentalizing."[5] Barring developmental disorders, adult humans mindread constantly. We use eye gaze, facial expressions, body language, tone of voice, and the content of speech to try to determine what others see, want, think, and feel and to anticipate or explain their behavior on this basis. Some evidence exists that a few other species may engage in a degree of mindreading, with chimpanzees and bonobos (pigmy chimpanzees) being the top candidates. However, humans are almost certainly unique in that they mindread about mindreading: humans think about what others might be thinking about others' thoughts. Julia may wonder whether Ali thinks that Madeleine likes zorbing. This ability to think about thoughts, or meta-represent, also means that humans can reflect on their own thoughts: "I want to go to the party, but why? Is that something I *should* want?"

---

[5]Baron-Cohen, 1995.

Sophisticated forms of mindreading that humans use all the time are "joint attention" and "joint intentionality." Joint attention is when two individuals both attend to the same object and to the fact that they are attending to the same object. If an adult says, "Wow! Look at that!" to an infant, the infant may track the eye gaze of the adult to make a guess about what the adult is attending to and then check back to the adult to be sure they are both attending to the same thing. Joint attention may be critical for learning the meaning of words, but it is also the bedrock of cooperation. I often need to know that we are attending to the same thing before we can cooperate concerning it. Joint intentionality builds on this bedrock. Joint intentionality refers to the ability of two (or more) individuals to form joint goals that will motivate their cooperative activities while maintaining individual intentions. For instance, a three-year-old may know that for us to get candy out of a jar, she needs me to hold the jar while she turns the lid. Our joint intention is to get candy and her individual intention within that joint intention is to turn the lid but only when I am holding the jar still. This joint intentionality enables a kind of full-blown cooperation involving collective goals but specific roles within the cooperative effort, a trait that appears to be absent in any other animal.[6]

As mentioned in the previous chapter, human eyes—with unusually large whites of the eyes relative to the colored center—are rich sources of information we use to "read" each other's minds. The prominent contrast between the white sclera and the colored iris helps an observer determine the angle that eyes are directed and, hence, what someone is attending to. Subtle signs of emotional states and attitudes surround our eyes through eyebrow movement, little wrinkles at the edges, and even tears. Combining these emotional state signs with information about what someone is attending to gives us the ability to make inferences about what someone is excited about, scared of, interested in, happy to see, and so on. Psychologist Simon Baron-Cohen's work on how

---

[6]For a discussion of joint attention, intentionality, and related concepts, see Moll, Richter, Carpenter, and Tomasello (2008) and Tomasello (2019).

mindreading develops in human children places great emphasis on the importance of reading the mind through the eyes.[7]

As we continue to see, human social intelligence requires considerable brainpower. Humans learn, store, and retrieve information about dozens or hundreds of individuals and use that information to predict and explain each other's actions. Furthermore, humans augment that background information by considering what others have seen or been told in a situation, which helps them determine what those others know (as in the location of a hidden snack), and that information is used to refine predictions about behaviors. Humans constantly mindread. That big prefrontal cortex in human foreheads may have evolved because of its role in facilitating this sort of thinking. And yet human social intelligence is even more sophisticated, as we will see next.

Human social intelligence requires considerable brainpower.

Attention to social networks. Part of the background information about another individual that we humans know includes that person's network of friends, family, and perhaps enemies. We pay attention to how others feel about us and about others. We try to imagine who knows what about whom so we can navigate social encounters, get reliable information, and protect ourselves from unflattering information getting in the wrong hands. It does not take long for children to understand that offending or pleasing one person can mean offending or pleasing a number of others connected to that person and that sharing information with one person can mean that person is likely to share that information with others.

In small groups of individuals, this tracking of information concerning how people think about, feel about, and relate to others may not be too Herculean a feat. In an isolated group of five people, an individual can know plenty about each of the other four and something

[7]Moll, Richter, Carpenter, & Tomasello, 2008; Tomasello, 2019.

about how each of the other four feel about him or her, along with each of the others. Each person in this group would be keeping track of twenty different relationships: themselves in relation to each of the others, the others in relation to themselves, and the others' relationships in both directions with each other. Adding just one other person, a sixth, to the mix adds ten more relationships to consider. A group of ten would have ninety relationships to track. Now consider that humans traditionally have about 150 important people in their social networks. Humans, as a hypersocial species, require a very special brain indeed.[8]

If we consider a time when the ancestors of modern humans began to acquire the sort of social intelligence described above, we can suppose that a social intelligence "arms race" could have driven them to get socially savvy pretty quickly. Once they figured out that beliefs drive actions but beliefs can be false, these ancestors would have had additional information for determining who had reliable information and could be trusted—a definite aid when survival required the cooperation of multiple individuals—but they also could have tried to deceive each another. For instance, when it was time to share what everyone had hunted or gathered, ancestral Steve could have hidden some of it from everyone else so he could get a bigger share, knowing that others would not know he was cheating the group. The risk for Steve, however, would be getting caught and thereby ruining his reputation. If he became known as a cheater, others would be less likely to trust and cooperate with him. So Steve would need to be good at cheating, good at knowing how to protect his reputation, and good at knowing when the costs of getting caught and the probability of getting caught outweighed the potential rewards of cheating the group. Again, such social intelligence requires considerable brain-based computing power.

To catch cheaters and liars, these ancestors would have needed to improve in social intelligence. But as they became better at detecting

---

[8]Dunbar, Barrett, & Lycett, 2007.

cheaters, cheaters could have gotten better at cheating. They may even have realized that when they caught someone cheating, others would not necessarily know that that individual cheated, so this secret information would have become a resource that could be used to get cooperation from the cheater.

In addition to this conceptual arms race for learning how to cheat or detect cheaters in a social group, theorists have suggested that solving cooperative problems and more efficient teaching and sharing of information are among the factors that drove the evolution of better and better social intelligence.[9] By either sort of account (or their combination), it appears that living in large groups of specific individuals who could act on individually held information and motivations may have created pressure for more and more sophisticated social intelligence. This social intelligence was (and is) needed to track each other's trustworthiness and one's own reputation within a complex web of relationships that make up our social networks. While humans appear to be unique in the layers of social complexity that they automatically and almost instinctively navigate, shades of these dynamics do color some other social species.

For instance, some other social animals practice reciprocal altruism. That is, they do something for others now (such as share food) in expectation that those others will do something in return for them in the future. When wolves or lions hunt, not all of the individuals will catch something every time, so sharing is a common part of their social practice. What happens, however, if one member of the group is a freeloader and happily shares in others' food but will not share the food it procures? Some animals keep track of whether others have a good track record of reciprocating in this tit-for-tat fashion and cooperate only with good reciprocators.[10] Humans do this, too, but our large group sizes mean it is much easier (in theory) for freeloaders or cheaters to take advantage of the

---

[9]For instance, see Henrich (2015), Laland (2017), and Tomasello (2019). This topic is taken up in the next chapter.
[10]Wilkinson, 1988.

kindness of others. Hence, we may expend mental energy tracking who is trustworthy and trying to protect our own relationships.

Perhaps even more important than maintaining one's own reputation and discouraging cheaters may be the role social intelligence plays in gathering, sharing, and using high-quality information for solving such problems as determining what to eat, how to prepare it, where to go for water in the event of a drought, how to build a shelter, and so on.[11] Being able to acquire and share (often via teaching) high-quality information with others means each generation does not have to start from scratch in navigating life's demands. Any new insights can build on what has been passed down. In this way, our social intelligence enables cumulative cultural innovation, a point that receives more attention in the next chapter.

Likewise, mindreading and social intelligence could also be valuable for cooperation and care. Knowing particulars about how you build a shelter and what you might be thinking or intending at any given moment gives me ways to anticipate how I can help, facilitating our work together. If my mindreading gives me a good sense of what you desire, I can anticipate how to meet those desires. Consider a common form of family cooperation, preparing and sharing food. Members of a family have different nutritional needs, dietary restrictions, and even food preferences. Tracking each of those individuals' needs and wants and how they might be different today than yesterday (e.g., Skylar is going to be working hard in the forest for most of the day, so he'll need a heartier breakfast than normal) would enable greater care for each other. And so, care and cooperation could also have served to encourage the evolution of social intelligence.[12]

SOCIAL EMOTIONS AND BONDING

Living in large groups requires a lot of information tracking, but human sociality is not just about keeping track of individuals and who knows

---

[11]Laland, 2017.

[12]Many recent authors have stressed the cooperative nature of humans and not just competing with or managing each other. Such scholars include Boyd (2018), Fuentes (2019), and Tomasello (2019).

what about whom. Gossip is not the only game in town. The individuals we spend so much time with and invest in are dear to us. They conjure greater degrees of empathy and loyalty in us. We become emotionally bonded to each other as a result of natural mechanisms, including the attachment mechanism and various social grooming strategies.

Attachment. Ducklings and chicks imprint on their mothers; humans attach.[13] Humans form emotionally rich bonds with their caregivers that subsequently color their relationships throughout life. A secure attachment is characterized by trusting Mom's devotion enough to be able to explore and play without Mom physically present at all times. In contrast, an anxious attachment is clingy and fearful of Mom leaving for even brief periods. An avoidant attachment is so absent of trust that the child just avoids or ignores Mom. Psychologists think these characteristic patterns of attachment are generated by an evolved mechanism that weighs the responsiveness of parents early in a child's life, which subsequently generates a social attachment style that is adaptive for the child. If a parent is capricious and violent, it may be best for the child to avoid getting too close. If a parent constantly shows signs that the world is full of dangers, then clinging anxiously may be a good survival strategy. If, however, a parent is predictable, confident, and comforting, then using that parent as a safe base for increasingly independent explorations of the world may be most adaptive. Human infants and children are so slow-maturing and, hence, vulnerable to threats in the environment that they need these adaptive routines to help keep them from danger.

Social grooming. Relationships with close family members, best friends, and spouses show a high degree of trust and emotional connection, but some relationships are based on looser bonds, such as those we enjoy with good friends. Nevertheless, even this looser bonding is a product of feelings of trust and connectedness, not just of knowing information about the other. These feelings are cultivated by time spent

---

[13]Bowlby, 1982.

in social grooming, engaging in activities that help elicit trust and positive feelings toward others through the release of endorphins.[14]

Monkeys and apes are famous for nitpicking and other types of social grooming. We usually think they pull bugs out of each other's fur to do each other a favor or procure a little snack, and that may be accurate, but an important product of this activity is increased bonding. This kind of physical contact releases endorphin hormones that give pleasure and help bond individual animals to each other. Though humans do not (usually) pick creepy-crawlies out of each other's hair, we do socially groom each other. For example, watch schoolgirls interact. They brush and braid and stroke each other's hair. Humans walk arm in arm, hand in hand. We humans hug and kiss and caress each other. These activities trigger the release of endorphins, thereby bonding us with each other.

Social grooming takes time away from other survival activities such as food gathering and preparation. Consequently, the number of others that an animal can socially groom is limited by hours in the day as well as physical proximity. Humans, however, have much larger social groups than other primates. How do humans "socially groom" all of these people? We humans have some very efficient forms of social grooming that allow us to strengthen our bond with multiple individuals at once.[15] In addition to being in direct physical contact, humans can evoke endorphins by laughing together. Though chimpanzees may engage in a kind of panting protolaugh, humans are the only animals that enjoy full-on belly laughs. Laughter is good medicine because it releases pleasurable hormones. Laughing with others, then, helps us feel good about those others and comfortable with them. Research has also shown that synchronized or coordinated activities produce feelings of trust and security with others, likely because of endorphins being released analogous to social grooming. Rhythmic, synchronized drumming, clapping, dancing, and rituals are all ways that humans can "socially groom" large

---

[14]For more detail on social grooming in humans and nonhumans, see Dunbar (1998).

[15]Dunbar, 2004.

numbers of others simultaneously and thereby build communities of trust.[16] These activities are all distinctively human behaviors that traditional societies incorporate in celebrations and parties as well as religious and cultural observances.

Note that physical proximity and especially touch seem to be very important for social grooming, whether in humans or other apes. In contemporary conditions, we may use electronic communication technologies to help us maintain relationships, and surely they can help do so to a certain extent. But the very physical nature of social grooming and the primacy of touch in triggering neurochemical cascades that build closeness mean the deep bonds of trust and positive regard produced as a result are hard (if not impossible) to mimic virtually.[17] Digital social networking is no substitute for in-person social networking.

SOCIAL NETWORKS

Our insatiable collecting of social information and our social grooming of others result in predictably structured social networks that can be likened to concentric circles of intimacy as in a target or dartboard. The typical numbers of people in these rings are, beginning with the center and moving outward, five, fifteen, fifty, and 150.[18] The bull's-eye of a typical person's social network is about five best friends or family members with whom that person has very high levels of emotional intimacy and trust. Around those five is another eight to ten good friends or close family members for a total of about fifteen. The next ring consists of the previous fifteen plus roughly another thirty-five close relationships with people one is content to see less frequently than the first fifteen—these would not necessarily be the go-to people in a crisis but are still trusted, valued, and cared for. Around these fifty people average another one hundred personal relationships: people one knows personally (not just professionally)

[16]Weinstein et al., 2016; Whitehouse & Kavanagh, 2021; Xygalatas, 2021.

[17]Virtual touch technologies are being developed, but it is yet to be seen whether they can fully duplicate social grooming.

[18]For details on social networks, see Hill and Dunbar (2003).

and with whom one feels a degree of emotional closeness and trust but to whom one does not reach out quite as frequently or feel quite as intimate with as the inner fifty. In total, then, an average active human social network consists of approximately 150 people. Outside of that network are acquaintances and nonpersonal relationships. Not coincidentally, when one looks at the sizes of traditional villages, churches, and even military units, social groups tend to average around 150 individuals. Anthropologist Robin Dunbar has argued that this social group size is the largest that can self-regulate through social bonds. Everyone knows each other and can keep track of who is contributing and who is not. Beyond this size, power structures and governance may become necessary.

Traditionally, when settlements became larger than 150, personal social networks of around 150 people consisted of those who lived in proximity, including extended family and friends. If we assume the average family had four children who reached adulthood, then the typical young adult would have had two parents, three siblings and their spouses, twelve aunts and uncles (if all were alive and married), sixteen cousins and their spouses, and possibly four grandparents and great-aunts and great-uncles. Then there would have been nieces, nephews, cousins' children, and one's own children. Some may have moved to neighboring villages or further afield, but easily one-third or more of the 150 personal relationships would have been family members by blood or marriage. The rest of the network would have been almost entirely friends and neighbors. Given the relative lack of mobility of populations up until the last two hundred years, it would have been a rare occasion for most people on earth to meet someone with whom they did not have a friend or family member in common. Almost everyone would have been a familiar face known by you or someone you knew, so making a completely new friend after childhood would have been very rare.

An adult who grew up in a nontransient town of fifteen hundred people would likely have had a personal relationship with about one-tenth of the citizenry, be acquainted with a third, and recognize almost everyone. It would have been very easy to determine who was a friend or relative of whom and those with whom one had social bonds and obligations. Keeping secrets, particularly of moral failings, would have been very difficult, so one would have had to carefully guard one's reputation, especially since relocating to another community would not have been an easy option. In this way, even towns of more than 150 individuals could still have bred a high degree of self-regulation through social connections. The greater the number of people living in close proximity, however, the greater the opportunity for anonymity and ability to get away with antisocial behavior—a point we will return to in chapter six.

THE SOCIAL GAP

Human nature is social. Humans automatically track what other people are paying attention to, desiring, thinking, and feeling in part by watching their faces—especially their eyes. We humans predict and explain each other's behavior in terms of the mental states that we mindread. This social nature is tuned to about 150 personal relationships, a large portion of which are relatives.[19] Our nature equips us very well for social conditions in which we recognize essentially everyone and can socially place them in terms of their relations, social status, and obligations.

Humans, however, increasingly live in a social environment radically different from what their social brains are naturally equipped to handle. In the last decade, urban dwellers have outnumbered rural people worldwide for the first time in human history.[20] Most humans can no longer expect to go through each day never seeing a complete stranger. Rather, urban humans can expect to know only a small fraction of the people they see daily. Anonymity is the norm.

---

[19]As part of an argument against kinship being the explanation of large-scale cooperation in humans, Boyd (2018) stresses the *low* average degree of relatedness in human groups (pp. 79-80). Nonetheless, his reported estimates are consistent with the hypothetical I have given here. If two-thirds of one's network are not related and a portion of the "relatives" are relatives only by marriage, the average degree of relatedness of such a network would be around the 0.05 level that Boyd suggests is supported by empirical and theoretical work looking at relatedness in human hunter-gatherers.

[20]United Nations, 2014.

Scholars who research human thriving commonly regard loving and caring for others and contributing to the general social good as key features of a thriving life.[21] Natural capacities for caring for others, however, appear to be impeded by contemporary social niches. One example is the tendency for a large number of bystanders to simply watch or ignore when an obvious travesty takes place. Perhaps the most notorious example was the hit-and-run death of two-year-old Wang Yue in Foshan, Guangdong, China, in October 2011. Caught on video and broadcast on the internet, some eighteen people simply walked around the injured child instead of helping. Though this incident became a symbol of alleged Chinese social and moral apathy in a country famed for being "collectivist," the fact is that in largely anonymous urban settings, people often ignore the plights of others, whether in China, the United States, Brazil, Germany, or Nigeria. One way to understand why humans behave in this antisocial, uncaring manner is to compare our natural social cognition with the current social niche.

Humans increasingly live in a social environment radically different from what their social brains are naturally equipped to handle.

Traditionally, we would likely have recognized anyone who was injured or in need of help, and they would have recognized us. Simply being known helps spur us to action. People are much more likely to behave antisocially when their identity is hidden (as when they wear masks or uniforms).[22] Furthermore, in a traditional community, odds are that even if you don't personally know the person in need, you know someone who does.

Interestingly, it is one of our social instincts, now placed in historically unusual circumstances, that leads to bystander apathy and other instances of failing to help others.[23] Humans have a strong tendency to

---

[21]Lerner, 2007; Bundick, Yeager, King, & Damon, 2010; King & Mangan, in press.
[22]Rehm, Steinleiner, & Lilli, 1987; Diener, 1979.
[23]Latané & Darley, 1969.

socially conform, to go along with what most people are doing. In many settings, this means we are cooperative and relatively docile, which are generally positive features. So in the small-scaled societies that characterized most of human history, if a child fell down a hole, for instance, that child's siblings, friends, and parents would rush to the child's aid, and then mere acquaintances would follow because of social conformity pressure. Now, in larger mostly anonymous societies, when we humans see so many other people failing to act, our tendency toward conformity encourages us to likewise avoid action. It is as if we unconsciously reason, "If no one else thinks help is needed, then it must not be," or "If no one else is acting, there must be good reason not to help; maybe it is dangerous to do so." The combination of anonymity and social conformity leads to dramatic failures to help each other.

We may want others to catch a vision for positively changing our community or nation, but note how unnatural this vision may be in current environments. If our social brains are better suited for communities largely composed of friends, family, and acquaintances, then wanting to serve one's community has considerable natural impetus. Not only would each member know a lot about their community's needs, strengths, and challenges, each member would naturally care for at least a large proportion of the individuals in it. Empathy is easiest when its target is someone most naturally regarded as similar to oneself: relatives, friends, and those who have similar customs and interests. Now, however, "your community" can be code for thousands or tens of thousands of anonymous others grouped together only because they live in the same general vicinity. It is much harder to care self-sacrificially for such abstract, temporary, and arbitrary groupings of people.

We humans are arguably more self-interested by nature than perhaps we should be. Life requires us to balance self-interest with the interest of those around us—particularly our family and social network—and too often self-interest outweighs altruism. The gap between contemporary urban social niches and our natural sociality makes the situation even worse. Research suggests that we are relatively trusting and cooperative with people with whom we have engaged in social grooming, whether through touch, synchronized actions, or even participation in common religious rituals.[24] The number of anonymous and semi-anonymous humans with whom urbanites interact every day virtually guarantees they will socially groom no more than a tiny fraction. Feeling any kind of emotional bond that leads to trust and cooperation with the people with whom one interacts each day is difficult to cultivate in these conditions.

Feeling any kind of emotional bond that leads to trust and cooperation with the people with whom one interacts each day is difficult to cultivate in these conditions.

The social gap creates additional specific liabilities for children and youth. A good predictor of a teen with good school outcomes who avoids illicit drugs and alcohol, exhibits good citizenship, cares for others, and otherwise shows signs of thriving is whether adults personally invest in that teen's life. About five adults mentoring, taking interest in, supporting, and otherwise caring for a teen seems to be a good benchmark, and even better still is when those five adults know each other and know that they are invested in the same teen.[25] This network closure—when the adult champions of a teen know each other—helps the adults coordinate their support of the teen and also communicates to the teen that he or she has a safety (and accountability) net.[26] Notice that this magic threshold of five invested adults who all know each other would have been commonplace in our ancestral past. The typical teenager would have had two parents, up to four grandparents, a half-dozen or so aunts and uncles, and maybe even some older cousins and family friends all caring for him or her. Today's smaller, scattered families and transient lifestyles combine to deprive teens of the natural adult support they need to thrive. Consequently, we often have to intentionally socially engineer

[24]For instance, see Cohen, Ejsmond-Frey, Knight, and Dunbar (2010) and Sosis and Ruffle (2003).
[25]Damon, 1997; Benson, 2006.
[26]Smith, 2003.

relationships so teens have a resource that would have been an ordinary part of their niche in the past. Humans now have to put forth more effort and be more creative to overcome the social gap we have created. As a result, those with the social and economic capital and the time required to curate such a portfolio of relationships for their offspring have a distinct advantage in our contemporary niche over those who have less access to time, economic, and social resources. Many contemporary urban environments, then, require more social engineering to address the social-relational gap created by these socially engineered environments.

5

GETTING AND USING INFORMATION

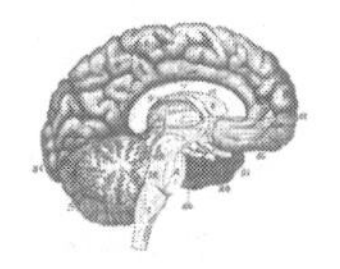

When we watch quiz shows such as *Jeopardy!* or play games like Trivial Pursuit, part of the fun is testing our own knowledge in comparison with others' and marveling at what others know. We live in a world in which one person can know a lot about gardening, another about cooking, another about Greek philosophy, and still another about twentieth-century African literature. Knowledge is very uneven in the world of contemporary humans in part because of the varieties of educational systems and the radically different niches humans face around the globe.

One thing that makes us human is our amazing ability to learn vast amounts of information from each other and use that information to address the demands and opportunities presented by our local niche. Part of human thriving, then, is using that ability effectively. Our long developmental period, coupled with the ability to build up cultural knowledge and learn it from one another, gives us far more opportunity to acquire knowledge and insight than any other species on earth. These peculiarities likewise give us considerable variability in what each human knows or does not know. This ability to learn niche-specific information enables humans to adapt to new challenges, divide labor, and become experts or specialists at certain survival problems. Anthropologist Joe Henrich goes so far as to suggest that it is not greater general intelligence that sets humans apart from the other great apes but, specifically, our ability to learn from each other.[1] In a similar vein, biologist Kevin Laland writes, "Humanity's success is sometimes accredited to our cleverness, but culture is what actually makes us smart." For Laland, "culture" means "the extensive accumulation of stored, learned knowledge, and iterative improvements in technology over time."[2]

This flipping of conventional wisdom by Laland and Henrich is well-argued. The basic claim is that little steps down the road of accumulating and sharing information and techniques for solving problems changed selection pressures on humans such that greater ability to acquire, share, and use such information was fitness-enhancing. This process may have led to some genetic changes, but the most dramatic changes likely had more to do with the social and cultural environments in which humans began to find themselves. These environments, in turn, made us "smarter."

It would be a mistake, however, to think that the only difference between humans and other apes (or even elephants and dolphins) is the richer social and cultural environments humans grow up in and that otherwise we would have problem-solving and other cognitive skills similar to these animals. Even if one is tempted to read Laland as advocating such a view, Henrich's and Michael Tomasello's reviews of the distinctively human cognitive capacities that are now part of human nature should put to rest any such view that humans are merely ordinary apes that happen to grow up in better educational environments.[3] Some of these tools in the human conceptual kit were described in the previous chapter and more are sketched below.

---

[1]Henrich, 2015, especially chap. 2.

[2]Laland, 2017, p. 7.

[3]Henrich, 2015; Tomasello, 2019.

Nonhuman animals are generalists when it comes to life skills and knowledge. Each one has basically the same survival toolkit as others of the same species. We can imagine a time in which a single human may have had skill in all of life's basic demands, such as making and using fishing tools, building a fire, cooking food, building a shelter, making basic clothing, and parenting and caring for children. But even that imaginary past is probably largely a fiction. Humans, as far back as the archeological record allows us to glimpse, are a species of cooperating specialists. For thousands of years a single human could become an expert toolmaker, house builder, clothing maker, hunter, farmer, or cook, building on the expertise of those who previously acquired skills in those roles.

Having societies with built-in divisions of labor and specializations of tasks has meant that insights could accumulate, technologies could build on previous discoveries, and more complex practices could be developed. But ever-ratcheting-up specialization also means that each succeeding generation has to learn one or more of those specializations. With larger group living and greater specialization, more and more options for expertise acquisition have become available and required of each subsequent generation. As these skills are based on the insights of generations of innovators who came before, much of this learning requires considerably more time to acquire. Learning to repair a car takes more time than learning to repair a spear, and learning to build a contemporary house takes many more years of training than learning to build a tent. Increasing transience and technological developments have created greater changes in environmental demands from one generation to the next, resulting in even more required learning of each subsequent generation. The amount that contemporary urban children in the developed world have to learn in order to earn gainful employment is staggering, often taking

More and more options for expertise acquisition have become available and required of each subsequent generation.

upwards of sixteen years of formal schooling after five years of preschool life. By the time many contemporary humans are ready to start their adult life, the typical monkey is a grandparent facing its twilight years.

Deliberately acquiring and using information is something humans do naturally, but the conditions under which humans optimally learn are not unbounded. Psychologists and anthropologists have studied the most natural ways for humans to learn, starting with from whom we most naturally learn.

SOCIAL LEARNING BIASES

Because we humans are immensely social, we are constantly surrounded by potential sources of information and people from whom we could learn. These other people serve as sources for acquiring expertise by modeling how to be a good person and how to navigate social and survival challenges. With so many potential models around, however, it may be difficult to select whom to turn to for sound information and advice, whom to imitate and regard as a role model, and whom to trust. To address this challenge, it has served humans well to have unconscious heuristics or biases[4] for picking favorite models, teachers, and mentors—those people to whom particular attention will be paid and credibility granted.[5] I sketch four of these important social learning biases below: prestige, skill, conformity, and similarity. I wish to be clear, however, that though these may be natural biases of social learning, it does not mean they are always most useful or desirable.[6]

---

[4]Here and throughout I use the term *bias* to refer to tendencies or predilections to favor one way of thinking over others, as is common in the psychological literature. Often in common usage a "bias" is a bad thing because it suggests prejudice, but I do not mean to suggest such a connotation. Of course, some biases may lead to errors in thinking and acting, but some may also be useful for solving complex problems efficiently.

[5]Henrich, 2015; Henrich & Broesch, 2011. These are sometimes called "context biases" to contrast with "content biases."

[6]Indeed, to anticipate chapters seven and eight, I think Christian anthropology is a helpful corrective to what may be natural tendencies to disregard the positive example of others simply on the basis that they are not prestigious or do not appear similar to us in superficial ways. Christian anthropology affirms that all of us are God's beloved creatures and can share kinship with God's chosen perfect image, Jesus Christ. So regardless of status, prestige, sex, age, or social group (including race and ethnicity), we have a deep commonality that pulls us together as humans.

Prestige. Why do we pay attention to the opinions of celebrities on everything from religious and political issues to what to buy, eat, and wear? How does athletic prowess or acting ability make one an expert on the environment, parenting, or geopolitics? As celebrity culture amply demonstrates, humans gravitate to prestige. Prestigious people serve as models of what it means to succeed in a particular society. High social standing is marked in different ways in different cultures, but humans quickly pick out those of high prestige and selectively attend to what they think and do. Furthermore, research shows that we tend to like people who act like us, so acting like prestigious people—such as demonstrating similar tastes, preferences, and manners—encourages them to bestow favor.[7]

It has served humans well to have unconscious heuristics or biases for picking favorite models, teachers, and mentors.

One enduring marker of prestige is being well-known. Someone who is mentioned repeatedly in conversation with positive tones, even when not present, is likely to be considered a prestigious person, at least historically. Today, with mass media, someone can become famous without any of the traditional markers of prestige and certainly without any relevant skills or knowledge. Being widely followed on Instagram or having a popular YouTube channel for makeup tutorials or covering popular songs is enough to make one a celebrity, even if much greater skills in these areas are present in the people around us. Nevertheless, our human prestige detectors sound the alarm when we see a celebrity and, whether we want to or not, we unconsciously take their testimony and example more seriously than that of other people. These psychological dynamics are why famous athletes or performers can sell us products completely unrelated to their field of expertise, such as clothing, food, and even jewelry. This gravitation toward prestigious role models may also be a

[7]Henrich & McElreath, 2007.

residue of our ancestral past in which high prestige also meant high skill or physical strength. Under these traditional conditions, imitating this person would have been a good strategy for building up one's own skills, thereby improving fitness.

Skill. A sensible bias, even in our day, is to attend to those who have skill or success in the area we wish to learn about. If one wants to learn to cook well and knows some skilled cooks, learning from them is a pretty good bet. To be a great parent, copy the strategies of a great parent. From childhood, we humans automatically try to copy and learn from those we perceive as skillful. Of course, one way to determine who is skillful is to consider how famous they are for their skillfulness. If this skill is in a valued area, prestige may result. Prestige bias and skill bias, then, may point in the same direction, reinforcing one another. Sometimes determining who is skillful is easy: a skillful hunter brings home more meat; a skillful homebuilder's houses withstand storms better. But in many domains, perhaps especially in today's complex societies in which outcomes less directly follow from one's skills, it may be hard to pick out who is genuinely skillful. Prestige, then, can serve as a proxy indicator of skill. Alternatively, we can rely on the collective wisdom of others around us. Even if I can't pick out who is skillful, perhaps my community knows.

Conformity. "But, Mom, everyone is going to be there!" As social animals with something of a herd mentality, humans show great sensitivity to what the majority of people—particularly people relevantly similar to oneself (see below)—seem to be doing. Hence, if the crowd is paying particular attention to something someone is saying, the speaker immediately gains the benefit of the doubt. The additional credibility someone may gain from their skill or prestige can really snowball when a critical mass of people starts listening to or imitating that person. This conformity bias may lead to disaster, as in the case of evil or misguided charismatic leaders who gain a large following, but by and large, looking to an individual seen as credible by everyone else is a reasonable heuristic for picking out a useful information source or role model. In times past, it would have been difficult for the village idiot to acquire a following.[8]

Even when not aimed at determining who in particular to listen to and copy, our strong tendencies to conform help us gravitate toward ideas and practices shared by the bulk of people in our community. And while strong conformity may seem like it endangers innovation and new, better solutions, it can also help prevent unproven and possibly dangerous ideas from taking hold. Conformity can also stabilize and consolidate the conventional wisdom of a group so that there is something to build on. Theory and mathematical modeling of decision-making support the idea that in situations in which one's own judgment cannot be trusted and the environment is not changing too quickly, it is often a better strategy to go with convention than try to innovate.[9]

Similarity. When election season rolls around, candidates at all levels of government try to demonstrate how much they are like "regular folks." The candidates understand that we tend to like people who are like us, a tendency that begins in childhood. Research even suggests that we find people who look a little bit like ourselves more attractive![10]

Not only do we tend to like people who are like us, we humans also tend to look to others like ourselves both for information and as role models. Therefore, prestige, skill, and conformity biases are tempered by a similarity bias. We will imitate prestigious and skilled people with whom we identify, who we think are like us, and whom other people like us are flocking toward.

Of course, similarity can be found along other dimensions, but three crossculturally important ones are sex, social group (often including

---

[8]Mass media and the ability for image to be massaged by expert storytellers and propagandists have changed this possibility. For a thoughtful and entertaining critique of how someone with no particular relevant skills or gifts could become a public figure with a large following, I recommend the 1979 comedy-drama *Being There*, starring Peter Sellers.

[9]Boyd, 2018, chap. 1. If Boyd is right, we may be facing an interesting moment in some contemporary societies. The tendency for our species to focus on learning from the group instead of innovating—arguably a psychological inclination that helped us accumulate so much cultural knowledge yielding so much technological complexity—may be becoming less adaptive because human environments are beginning to change rapidly. Perhaps we are on the cusp of yet another nature-niche gap.

[10]Fraley & Marks, 2010.

ethnicity), and age.[11] We humans tend to imitate others of the same sex: boys look to other males and girls look to other females as role models and sources of credible information. Likewise, when given the choice, humans turn to whomever they consider their own people: their own clan, tribe, social group, or ethnicity. While such a tendency may turn into racism, it need not. Actually, when children are asked whom they trust to give new information about the name of an object, it is not others of their race they gravitate toward but people who speak like them, who have similar accents, and who use language similarly—markers of a common in-group.[12] Considering that three hundred years ago (or even less for most people) a human of one "race" almost never came into contact with someone of another race, humans would not have needed a special sensitivity to physical markers of race as determining who was part of "my people" but would be more sensitive to culturally variable factors. For instance, accent and manner of speech can vary from one village to another, even if the population looks similar. A potential outsider or enemy could have been found out by their funny speech. From infancy we use accent as a determinant of who is worth listening to.[13]

Age is another important dimension that humans use to determine how similar others are. It is almost a cliché that the seemingly "cool" older kids have more sway over children's way of thinking than parents would like. From whom does a child learn the school social code, how to play games, or how to dress? Older kids. Research has suggested that children tend to gravitate to other children just a few years older as sources of credible information.[14] Five-year-olds tend to be interested in what the six- and seven-year-olds think and do. Twelve-year-olds find fourteen- and fifteen-year-olds fascinating. High schoolers mimic college students. This tendency may be adaptive for children because instead of trying to learn from adult models who vastly exceed their knowledge or skill, it

---

[11]Souza, Byers-Heinlein, & Poulin-Dubois, 2013

[12]Kinzler, Shutts, De Jesus, & Spelke, 2009.

[13]Kurzban, Tooby, & Cosmides, 2001.

[14]Henrich & Gil-White, 2001.

may be easier to learn from someone just a couple steps ahead of themselves. Furthermore, it may be that slightly older children better understand the learner's perspective than much older children or adults. Of course, if a slightly older kid is also prestigious and skillful, all the better.

CONTENT BIASES

Social learning biases impact from whom we learn, but another set of biases impact what we learn quickly and easily. These content biases are born out of our natural psychology mentioned in chapters two and three. Much as birds are biased to learn bird songs, particularly of their own species, humans are inclined to learn ideas and skills that resonate with their natural psychological systems.[15]

Since a thorough cataloging of content biases and their implications for human learning would be a major divergence, here I only observe that, all else being equal, humans will be more eager and find it easier to learn about those things that most closely relate to fitness problems of our human past. Because at least some aspects of our nature are relatively slow to change and most of human history was spent under very different environmental conditions from today's urban settings, a good rule of thumb is that if it resonates with a Stone Age mind, it will be more attractive to contemporary humans broadly. Why does the teen who seems unable to grasp the simplest ideas concerning algebra carry around encyclopedic knowledge of the latest school or celebrity gossip? Gossip focuses on information that is relevant to social survival and reproduction: who has wealth, is prestigious, is untrustworthy, is engaged in romantic liaisons with whom, and so on. Similarly, a woman who is unable (or does not care) to follow theories of macroeconomics may still be an adept local merchant, and a man who never could understand high school physics beyond simple mechanics may still be a skilled homebuilder.

[15]Barrett, 2012; McCauley, 2011. The comparison with bird songs comes from a quip biologist David Lahti once made at a workshop about how many songbirds seem to require hearing their elders sing in order to learn their song, and yet they only learn their species' song. The point is that specific bird species have content biases for social learning.

Building dwellings and exchanging goods and services are closer to pan-human survival problems than are theoretical and academic reflections on these problems.

If ideas most relevant to historically pervasive fitness demands are easier to acquire, then ideas and information that are relatively unnatural—foreign to these pervasive fitness demands and the way our minds naturally function—will be difficult to learn. Ancient insights will be easier to acquire (in general) than cutting-edge knowledge. For educators and policy-makers enthusiastic about STEM subjects (sciences, technology, engineering, and mathematics), this observation should be sobering. While having some toeholds in ordinary problem-solving and interest in the natural world, these subjects importantly build on relatively unnatural modes of thought that are foreign to our Stone Age minds.[16] The typical six-year-old will learn to dance faster than he or she learns algebra and animal care faster than genetics.

If it resonates with a Stone Age mind, it will be more attractive to contemporary humans broadly.

It may also be that some kinds of learning are most effective at particular points in development, and not just because learning builds on what came before in a curricular way. Developmental psychologists have repeatedly observed that many subsystems for learning have sensitive periods during development in humans as well as in other animals. That is, there are windows during which a little exposure to the material goes much further than it would have at other points in development.[17] The imprinting of chickens and ducks that I mentioned in chapter four is a good example. An adult duck will not imprint on another duck—the

---

[16]McCauley, 2011; Sperber, 1996.

[17]The cause of sensitive developmental periods is very much contested and likely varies depending on the domain. For instance, it could be that children seem to learn languages so much better than adults because they are less inhibited, have more flexible sound discrimination, and have less interference from other acquired information. A sensitive period need not imply a specialized, genetically coded cognitive device but could be a byproduct of various learning routines, dispositions, and social conditions.

sensitive period for imprinting is shortly after hatching. Analogously, the sensitive period for language learning is the first six years of life, with some lingering receptivity up until puberty, after which language-learning capacity drops considerably. Music and body control as in dance and sport may have early sensitive periods as well. In contrast, certain aspects of social intelligence and analytic reasoning may not even have a growth spurt until later childhood or adolescence. Of course, these observations may have practical implications for helping youth thrive, as explored below.

EDUCATIONAL NICHE GAP

Given how important the acquisition of cultural and specialist knowledge is in any human context, success at this task is surely a component of thriving. In the modern world, fitting into a niche well means learning and using a lot of information from others. In many societies around the world, much of this learning comes through schools and other modes of formal education. What implications, then, does an evolutionary psychology perspective have for educating young people? I tentatively suggest three lessons. First, educational systems should include older children instructing younger children. Second, adult mentors, when practical, need to be appropriately similar to the children mentees. Third, educational emphases should be selected based on developmental appropriateness and presented in terms that will be naturally motivating.

Children teaching children. If children naturally gravitate toward slightly older children of their same sex and social group, then why wouldn't we want children of one age playing a more prominent role in educating those of a slightly younger age? Not only would such a strategy play on the natural dispositions of the younger children, it would also provide a rehearsal of key ideas for the older children. Teaching is a great way to learn. Nine-year-olds could teach seven-year-olds how to read, do arithmetic, and play kickball. Eleven-year-olds could help eight-year-olds navigate social conflicts on the playground and learn to dance. Academic and cultural knowledge as well as social skills could be taught and reinforced in this way, as is done in some schools and less formal settings in many parts of the world.

Pam's children attended a kindergarten-through-eighth-grade school where such strategies were effectively employed and extremely well-received. The "K-Pal" program, which involves sixth-graders mentoring kindergarteners in reading, is one such example. For an entire year, sixth-grader and kindergartner pairs meet and read together weekly. These relationships become a memorable experience for the younger and older youth, and the kindergartners show increased interest and abilities in reading. The school has created various programs throughout its curriculum in which older students instruct, mentor, and lead younger students. These strategies are far different from the age-segregated, only-adults-teach-children model of learning so familiar in the Anglophone world that is actually a relatively recent human invention with deeper roots in industrialism than developmental psychology.

The right adults for the right children. Not all teaching is effectively delivered by slightly older peers. Children do need adults caring for and teaching them. Particularly, as children become teens, adults become the appropriate mentors for them. If the goal is for sixteen-year-olds to develop into effective, thriving adults, they should be socialized and educated by adults. Twenty-two-year-old role models may be a better fit than seventy-two-year-olds for most learning, but teens still need adult mentors. From the perspective of evolutionary psychology, I predict that teens will be most responsive to information from and the example of those mentors with whom they identify.[18] Same-sex mentors who are the sorts of people the teens imagine themselves becoming would be preferable teachers of what teens should know, what they should value, and how they should behave.

These observations suggest that there was something apt about the now old-fashioned apprenticeship and "work-study" models of education

[18]As noted above, "race" is not the most natural (or desired) operative category for identification. Bear in mind that in ages past, these "teachers" would often have been parents, extended family members, or friends of the family—people with high degrees of identification with the student.

for older adolescents. Sixteen- to eighteen-year-olds spending half or more of their day in a work environment with adults, receiving on-the-job training in a vocational area they intend to pursue, fits more naturally with how humans learn and develop than present-day high school environments. These mentoring relationships can pair adolescents with the sort of adults they want to be.

Jean Rhodes teamed up with the Search Institute on a study titled "You Get Me!"[19] Her research demonstrated that mentoring relationships characterized by showing both warmth and interest are more effective than randomly matched mentorships. When youth feel they are understood or share things in common, those relationships have more influence on them. In addition, mentorship relationships that focus on nurturing and equipping specific skills that are of interest or concern to youth are far more effective than general mentorships.[20] Given this tendency, mentorships around common interests can be excellent opportunities to expand the education and imagination of youth.

This more relational apprenticeship model does raise some challenges. Although it plays to the human social nature and can work with natural motivations to imitate those one aspires to be like, it may not afford the depth and breadth of intellectual knowledge now necessary to survive and thrive in many current environments. Nevertheless, an evolutionary psychology perspective suggests that educational contexts that leverage the social and learning proclivities of humans in general, and also appropriated for the needs of individuals specifically, would help bridge the learning gap humans chronically perpetuate.

Educational contexts that leverage the social and learning proclivities of humans in general, and also appropriated for the needs of individuals specifically, would help bridge the learning gap humans chronically perpetuate.

[19]Yoviene & Rhodes, 2016; see also Raposa, Ben, Olsho, and Rhodes (2019).
[20]Christensen, Hagler, Stams, Raposa, Burton, & Rhodes, 2020.

Developmental appropriateness of content. As sketched in this and preceding chapters, humans have natural proclivities to learn certain types of information. Other sorts of information that fit less naturally with human minds require special support to be acquired: direct instruction, clever teaching and learning techniques, lots of rehearsal and practice, and sufficient motivation. In general, the more natural areas of learning—those that bear on or are a byproduct of historically common fitness problems—come easily to children at younger ages. The less natural areas need to build up later and more slowly when reflective analytic skills and learning strategies are stronger. Relatively recent and unnatural areas of knowledge such as macroeconomics, calculus, and genetics are usually and sensibly introduced to children later than arts, practical life skills, basic numeracy, and basic language skills.

Helping children thrive may mean a reordering of our curricular sequencing. Some features of typical school curricula run counter to what we learn from evolutionary and developmental psychology. Waiting until children are twelve or thirteen to introduce them to a second language, for instance, fails to take advantage of their natural language-learning proclivities early in life. It is not surprising that the majority of children in these school systems show little retention of two to four years of second language education after they leave school. Two years of strong exposure when children are four and five will typically have a greater impact than comparable exposure when children are fourteen and fifteen. Similarly, investing in children's musical education, whether instrumental or vocal, will yield greater dividends when begun in early childhood. When it comes to the natural world, practical familiarity with plants and animals, their needs and their uses, and how to solve basic mechanical problems and the like fit with the proclivities of the young mind, whereas formal science—the theory, methods, and findings of chemistry, biology, physics, and geology—are better suited to older children and adolescents.

When it comes to character and moral development, it may be that some virtues are also more easily acquired at certain ages. For instance, some research suggests that interventions meant to encourage empathy, care, and kindness toward others are most effective during the early school years.[21] Self-control skills, however, could be much more difficult to cultivate until adolescence.[22]

Part of helping young people develop into thriving adults, then, may be a reexamination and reorganization of educational systems so they reduce the nature-niche gap relevant to how people learn. Placing the right teachers and mentors, including other children, into the lives of pupils and sequencing content in ways that resonate with natural learning systems in the minds of humans may prove helpful.

IMPORTANCE OF SPARKS

Positive youth development literature, particularly the work of Peter Benson and the Search Institute, has highlighted the importance of youth having "sparks": activities and interests that motivate and engage them. According to this research, if youth have identified a spark for themselves—much like a purpose—they are more likely to thrive. Having a handful of adults encouraging or supporting that spark may be even better for young people.[23]

Much as in the case of finding a vocation or personal purpose, discovering one's spark was probably much less difficult in ancestral conditions. Rather than selecting from among dozens of interests, including sports, academic subjects, extracurricular school-based activities and clubs, faith-based activities, jobs, and summer internships, previous generations of teens probably had only a handful of options available to them as potential sparks. Primary options for vocations may have come from what one's same-sex parent, close friend, or relative did for a living, so they may only have wondered, *Do I want to be a farmer or a carpenter?* or *A dairymaid or a seamstress?* If other sparks were even socially acceptable, they probably consisted of a limited number of arts or

[21]Santos, Chartier, Whalen, Chateau, & Boyd, 2011.
[22]Romer, Duckworth, Sznitman, & Park, 2010.
[23]Benson, 2008; Benson & Scales, 2009.

handicrafts that a master, close friend, or relative could teach. Importantly, in small-scale societies in which there was (and is) much greater vocational uniformity, the mark of a life well lived would not necessarily have been tied to one's livelihood. Tent-making, fishing, or basket-weaving wouldn't be where one would look for life satisfaction. Our larger, diverse, cosmopolitan societies and the presence of mass media may paralyze today's youth with too many options and inadvertently or even deliberately imbue some pursuits with more worth than justified.

Further complicating young people's discernment of their sparks, along with finding adult champions for those sparks, is that many options presented to young people are largely unrelated to deep-seated natural human motivations and interests. For instance, according to the Search Institute, when adolescent males in the United States list their sparks, the most common type listed is a sport.[24] Though physical games and sports are a relatively natural form of human expression, desiring to be an athlete or developing one's self-identity and worth around athletics is unnatural from the perspective offered here. That sports are so important in the identity of American males is the result of a very peculiar cultural niche. Investment in sports leads to enhanced fitness for only a tiny portion of people, particularly considering the other fitness-enhancing opportunities sacrificed for the sake of participating in organized sports. Imagine, for instance, if the same amount of time and resources invested in a teen playing four years of high school baseball, soccer, or wrestling was instead invested in volunteering to renovate and build homes for the poor, caring for the infirm, or learning life management skills such as preparing food or maintaining homes.

Rather than supporting youth in the sparks that their own peers or mass media have fed to them, caring adult mentors—including educators—should be encouraging youth to reflect on sparks that have deeper and more enduring resonance with historically common problems. How will I survive? How will I provide for my family? What

---

[24]Scales, Benson, & Roehlkepartain, 2011.

skills can I develop that will improve the lives of those around me? What kind of person should I be in my community? How do I honor God?

To summarize, humans are distinctive in their amazing capacity to learn specialized information to solve problems their particular niches throw at them. Some of this stretching is good and healthy. Acquiring expertise is part of how humans survive and thrive, but this expertise acquisition process is shaped by human nature concerning what we are good at learning and from whom we are good at learning it. Nonetheless, cultural systems and institutions may create demands on young people to learn more and differently than they are naturally well-equipped to do. Another nature-niche gap arises, and somehow we need to find a way to bridge that gap without continuing to widen it for future generations. I do not pretend to have a comprehensive plan for reworking educational systems, but I do encourage educators in various contexts—schools, churches, youth-serving programs, and businesses—to consider insights from evolutionary psychology in developing their curricula and pedagogical methods.

6

SELF-CONTROL

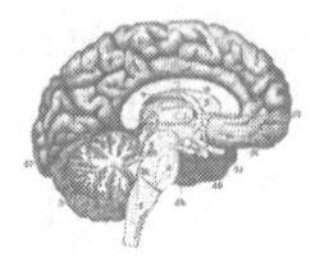

When I was eight years old, I was a child in perpetual fidgety motion. If I had been born twenty years later, I might have been diagnosed with attention deficit hyperactivity disorder. One day, when visiting my grandparents in California's gold country, I decided I wanted a "miner forty-niner" haircut from a barber in the historic gold rush town of Columbia. My mother relented, but with the warning that I must be very still in the chair so as not to ruin the haircut. I determined that I would remain still as a statue, so I consciously, deliberately mustered every ounce of self-control to not move. It became a game—one in which I succeeded! The barber heaped praise on me afterward: "I have never seen a boy stay so still during a haircut." Thereafter, my childhood self was still high-energy and squirmy, but I knew I could control my impulses if I made it a game with myself, and with each success, it became easier to win.

This deliberate regulation of one's thoughts and actions, what I call here "self-control," includes both willing oneself to act even when one does not want to do so and willing oneself not to act when one is tempted to do so. Self-control also involves the ability to weigh alternatives and willfully choose among them. All three of these kinds of effort may involve a special type of self-control that involves regulating one's emotional responses, or emotional regulation. As with sociality and the use and acquisition of information, we see shades of self-control in some animals. None, however, appears to exercise the degree of deliberate, intentional self-control that is a normal part of human life.

As with other animals, much human action is done "on autopilot" without conscious reflection or willful guidance. Nevertheless, we humans regularly consider how to get what we want, decide whether we ought to choose one option over another, and then choose to act. Sometimes these decisions about what we "want" or "ought" to do eventually become habitual and no longer require the same degree of conscious control, but that does not change the fact that the behavior once required considerable self-control to enact.

Consider learning to ride a bicycle. Initially, one has to think through how to mount the bicycle, how to move in a forward direction, and how to balance one's weight. One must push past the fear of falling. Eventually, these formerly conscious actions become one fluid result of a simple decision to ride the bicycle right now. An action that required considerable mental and physical control to learn becomes easy to manage without much conscious attention. In a similar way, many of our everyday modes of thinking and acting are built up from our distinctively human degree of self-control.

In this chapter we consider some of the domains in which we humans exercise self-control and reflect on how our niches may be changing in a way that creates challenges for the successful exercise of self-control, which in turn jeopardizes thriving. Self-control enables a number of virtues and moral behaviors that would be inconceivable without it. Choosing to pass up a small benefit now for a greater good later requires self-control. Restraining an impulse to lash out at someone requires self-control. Taking a risk for the sake of another often requires self-control. Inhibiting selfish impulses and putting others first requires self-control. Without the ability to pause, evaluate the

Inhibiting selfish impulses and putting others first requires self-control.

relative good versus bad of a proposed action, and then act, many moral actions would not be possible.

Being self-controlled all by itself gives one freedom and power to do many things someone lacking self-control could do, but those many things may include acts that are good, bad, or neutral. By itself, then, self-control is similar to great physical strength: aimed at the right ends it can be very good to have, but aimed at the wrong ends it is dangerous or vicious. Fortunately, the distinctive human ability to deliberately regulate our thoughts and actions is accompanied by a suite of other psychological tendencies that can be cultivated for virtuous action and thriving.

MORAL THOUGHT AND ACTION

Values and morality radically impact how people act. People will act in ways that may undermine their fitness for the sake of moral commitments.[1] Each day firefighters, police officers, and other emergency responders risk their lives for unrelated strangers in part because they are convinced that doing so is a morally right thing to do. People give large amounts of money to combat poverty and child slavery all over the world. They adopt children with severe disabilities and special needs. People donate, time, money, and blood to benefit strangers every day. Unlike the kin or reciprocal altruism we reviewed in chapter four, many of these types of activities do not bestow any documented fitness benefit on those who perform such acts—certainly not comparable in magnitude to the sacrifices—and yet this altruism is part of being human. However, not all moral behavior is about caring for other people. Some people regard care for animals, stewardship of the environment, personal character improvement, and a host of other efforts as moral decisions.

Some morally guided behaviors are still underexplained by current evolutionary perspectives, but progress is being made. An important, relatively recent step in the science of moral thought and behavior has been to recognize that moral decision-making is not always—perhaps

[1]William Damon and Anne Colby (2015) offer rich descriptions of the role of belief systems in historic moral heroes such as Dietrich Bonhoeffer, Eleanor Roosevelt, and Nelson Mandela.

not even typically—the product of reflective reasoning alone. Rather, we often experience intuitions or emotional reactions toward human actions as right or wrong even before we are able to consciously access reasons for these intuitions. This observation has led to the development of Jonathan Haidt's moral foundations theory.[2] In short, moral foundations theory argues that humans naturally possess a handful of emotional motivational engines that generate feelings and intuitions about the rightness or wrongness of certain classes of activities. These foundations are thought to be evolved mechanisms that have their origins in helping human populations solve basic problems of survival and reproduction in social contexts and may be culturally or personally overridden or elaborated into moral systems.[3]

Consider swiping your grandmother's false teeth while she sleeps, helping the opposing team beat your team in a sporting competition, or defecating on the altar in a place of worship. Without powerfully mitigating circumstances, most humans would instantly feel that these acts are wrong even before reasoning as to why they are wrong. Moral foundations theory says that many of our moral decisions arise from emotions and intuitions humans share across cultures. What, then, are these emotional, intuitive foundations for morality? The particular foundations are debated, but five proposed by Jonathan Haidt are worth considering.

Care/harm. The sights and sounds of someone in distress automatically trigger emotional responses in most people. Seeing someone harmed is repellant on a visceral level. These emotional responses are

---

[2]Haidt & Joseph, 2004; Haidt & Joseph, 2007. Critics of moral foundations theory point to its failure to capture all forms of cooperation and claim that it includes noncooperative domains (e.g., Curry, Chesters, & Van Lissa, 2019). Stressing morality as cooperation, anthropologists Oliver Curry, Daniel Mullins, and Harvey Whitehouse (2019) provide evidence for seven universal moral rules: help your group, support your family, return favors, be brave, defer to superiors, divide resources fairly, and respect people's property. In spite of this difference in emphasis, with Haidt's model, the key idea that moral thought has roots in multiple (not one) evolved mechanisms that generate emotions and intuitions on which explicit reflective reasoning and decision-making build seems to be a consensus position at the moment for those taking an evolutionary psychological approach.

[3]Using ancient scriptures and texts, C. S. Lewis argues for crossculturally recurrent moral themes, which he dubs the *Tao* (Lewis, 1943). Many of these overlap with the proposed "moral foundations." Like Lewis, Haidt scoured diverse wisdom literatures in his search for moral foundations.

argued to have arisen in our ancestors as part of the psychological package that facilitates intensely social living. Those people to whom a person has attached or formed an emotional bond through social grooming become his or her people or in-group. Because these intuitions are precultural, harming members of one's in-group (outside of special permission or circumstances) is immoral in almost any cultural environment. Failing to adequately respond to a member of one's group who is in distress also carries moral weight in most places.

Fairness/cheating. Try to serve different-sized pieces of cake to a group of children. Objections of "That's not fair!" will fill the air immediately. The fairness/cheating foundation may be a product of reciprocal altruism (discussed in chapter four). People may make sacrifices for others now in expectation that those others will return the favor later. Since such cooperative behavior relies on a common sense of what is fair or equitable, it may be that humans have evolved a sensitivity concerning who owes what to whom in social groups and whether or not someone is making the appropriate contribution for the amount of benefit they receive from other individuals or the group. As the divvying up of dessert illustrates, humans are sensitive to someone "cheating" others.[4]

Loyalty/betrayal. People intuitively sense the importance of being loyal to their core social group. Being part of a group carries certain obligations, and failing to meet those obligations is to be disloyal or, worse, a traitor. In most places, then, it is virtuous to sacrifice one's group and shameful to betray it. Our intuitions tell us that we carry moral obligations to the greater good. Family members will often put the needs of the family ahead of their personal needs. Citizens may sacrifice for their tribe or nation. Disloyalty may lead to expulsion because the disloyal person can no longer be trusted. Such intuitions may have been critical for human ancestors who began living in groups large enough that not everyone was so closely related that it was in one's best interest to take care of others at some cost to oneself (recall Hamilton's rule from chapter four).

---

[4]Cosmides, 1989.

Authority/subversion. Many raised in relatively egalitarian societies may mistakenly associate hierarchy and authority as being more characteristic of gorilla troops and lion prides than humans. We may all strive to be "leaders"—"alpha" males or females—and see nothing peculiar about that. Humans, however, have a long history of hierarchical social arrangements. Indeed, we are all born into one: the family unit. To mature properly, children need to respect the authority of parents. This non-egalitarian family arrangement was (and is) critical for children to survive and thrive. In times past, failing to properly respect parents and submit to their instruction could easily lead to lethal accidents, including getting lost, eating poisons, and being killed by deadly animals. Parents and other elders have acquired valuable knowledge particular to their children's niche that is critical for surviving and thriving. Even in suburban and urban areas, parents still steer their children away from dangerous streets, toxic substances, and unfriendly dogs. Similarly, when larger group coordination is needed, as in hunting a large and ferocious animal or fighting against an enemy army, hierarchy and respect of those in authority may be required for success. Failure to respect authority could lead to both individual and group calamity. Respecting authority, traditions, and the like may be such a fundamental component of success in group living that it has become part of the fabric of human psychology. Unless indoctrinated to the contrary, people generally recognize submission to and respect for authority, at least in some domains, as a moral duty.[5]

Sanctity/degradation. The idea of the "pure" and "sanctified" may derive from a psychological system that humans have for avoiding viruses, bacteria, and other pathogens. This avoidance is facilitated by psychological systems that strive to establish boundaries on clean or safe spaces and by giving us strong aversion to objects and spaces that are unclean

[5]As with many intuitions or "natural" social arrangements, their instantiations are not always good and sometimes very bad. Submission to misaimed authority can lead to travesties, as can sheeplike conformity to the movements of the crowd. The point here is not to valorize or demonize the submission to authority as a common practice but to draw attention to the fact that respect for authority may have deep roots in human history and psychology.

and, thus, potentially contaminating.[6] As omnivorous animals with a broad range of niches, we come into contact with lots of different foods and possible sources of illness. Hence, having a mechanism that makes us react strongly to these threats in our environment would be adaptive both ancestrally and currently. If we have not learned that something is okay to ingest early in life and it falls within a certain range of smells, textures, or appearances, we find it disgusting. This disgust reaction appears to be commonly harnessed by many cultural systems so that certain actions, or even people who have committed certain actions, may be seen as disgusting and potentially contaminating. These offenders are not just dangerous or bad but dirty, filthy, and defiling. Occasionally, such reasoning is extended to strangers, too, just by virtue of their being members of a particular group.[7] People also tend to mark off special sacred or ritually "clean" spaces in a way similar to how spaces involved in food preparation or personal hygiene can be carefully monitored for contamination.[8] Behaviors, people, places, and objects may all be categorized as contaminating or impure in an emotional, morally charged way. Sexual deviance is often characterized in this way. Sexual behaviors that fall outside group norms, such as pedophilia, polygamy, and incest, are often labeled as both wrong and disgusting. Similarly, emotional reactions may be mysteriously strong to outsiders when a holy space is "defiled" by a trespasser who is not ritually "pure."

These moral foundations appear to serve as key conceptual and emotional engines that drive much moral reasoning across cultures.[9] Particular foundations, however, are not necessarily equal in importance. For instance, whereas in most of the world strong family and community

---

[6]Rozin, Millman, & Nemeroff, 1986; Olatunji, Haidt, McKay, & David, 2008.

[7]Such a labeling may not be merely mean-spirited prejudice. In traditional societies, members of other people groups would have the potential of carrying diseases for which one's own immune system may be vulnerable. For indigenous peoples around the world, the age of European exploration and colonization was a time in which out-group members did bring "defiling" pathogens (Diamond, 2003).

[8]Contamination intuitions are commonly negative, but similar thought patterns may govern purification or positive contamination through ritualized behaviors (Boyer & Liénard, 2006).

[9]Haidt, Koller, & Dias, 1993; Shweder, Mahapatra, & Miller, 1987.

bonds lead to stronger affirmation of loyalty and ways of thinking about authority, these foundations may be relatively muted in more individualistic and libertarian Western cultures, perhaps through deliberate counter-indoctrination.[10] As Pam and colleagues have argued, moral intuitions may be integrated into a personal narrative that makes meaning of one's life and choices.[11] Think of it like building an edifice on the moral foundations. Principles for guiding one's life are like primary pillars and beams that rest on the foundational intuitions. These principles are joined together by narrative walls, flooring, and staircases that form a habitable identity. Religious systems and other philosophies one adopts may facilitate this habituation to the principles and integration of one's life with a narrative. Over time, living in this personal dwelling becomes easier, more familiar, and automatic, no longer burdening limited attention and willful self-control. Moral intuitions are not the whole story, then, but are foundational in important respects.

Another kind of variability we see across groups is scope of who counts as having moral relevance. Almost none of us regard a typical rock on the side of a road as relevant to our intuitions about who to harm, help, or be fair toward; almost all of us accept our parents and children as legitimate targets for care and fairness. Beyond those limiting cases, there is a lot of room for variability. Those who count as having moral relevance can range from one's kin group to all of humanity and even to some or all animals. Is it okay to harm people from outside the tribe? Perhaps. It may even be praised. But it is rarely permissible to harm one's family members or clan. Expanding the morally relevant in-group to large numbers of unknown others appears to be a product of only some cultural conditions. For example, in theory, Christianity extends high moral worth to all humans because all are

[10]Graham, Haidt, & Nosek, 2009. For an example of such efforts to overthrow the natural tendency to automatically respect authority, consider the perceived abuses of power by governmental authorities in the United States during the civil rights and Vietnam War era of the 1960s and '70s, which led to a common mantra of "Question authority." If not for the natural tendency to give authority the benefit of the doubt, at least much of the time, it may have been unnecessary to encourage people to question authority.

[11]King, 2020; King, Schnitker, & Houltberg, 2020; Schnitker, King, & Houltberg, 2019.

created in God's image and are potential "children of God"—an ideal that is difficult to achieve because it runs counter to other culturally enforced narratives about who constitutes the in-group.

An upshot of such an approach to moral thought as Haidt's is that if societally determined rights and wrongs align closely with one or more moral foundations, they will be easier to learn and act on. Indeed, we may feel impetus to do the "right" thing and avoid the "wrong" if right and wrong are close to the core of these moral foundations. In such situations, marshaling the willpower to do right may be fairly easy. But what happens when our moral intuitions are in competition with each other? Or if what we are taught is right is not closely related to one of these foundations? Those situations call for a strong dose of self-control.

SELF-CONTROL AS MUSCLE

Of course, possessing moral intuitions and being able to reason about what is right or wrong does not mean one will act accordingly. We often experience drives and motives that run contrary to what we believe is right. We do things we wish we had not, and not just because we got caught. Why, then, do humans do things we believe to be bad and fail to do what we think is right? A big part of the reason may be a failure of self-control. Sometimes it is difficult to muster the will to do what is right. Sufficient self-control is unavailable.

Recent psychological research reveals the utility of thinking about self-control as akin to an all-purpose muscle.[12] If Reuben attempts to lift a heavy load for the first time, he may fail simply due to lack of strength, but muscles get stronger through exercise, and after lifting lighter weights repeatedly, Reuben's muscles will build up strength over time. Then, with greater muscular strength, Reuben will be able to lift the weight that was previously too heavy. Similarly, self-control can be built up through exercising it. Failure to exercise it, however, can lead to weak self-control and failure in the face of even small challenges.

---

[12]Baumeister, Vohs, & Tice, 2007.

A strong self-control "muscle" would be an asset for many exercises of morality and virtue, ranging from courage to patience, but how does one build and keep strong self-control? Consider, again, building muscle strength. If you tried to build your muscles by lifting cars and buildings, you would probably make few gains. Why? The weight is too great for you to have any success. Likewise, losing self-control repeatedly does not build it up but may create a sense of helplessness. Self-control is exercised through trials, temptations, and decision-making through which one can experience success if one tries hard enough. However, unlike muscles, each uncontested self-control failure can undermine confidence and motivation to exert the effort in the future. If a temptation seems too hard to resist, we may not even try. What we seem to need, then, is a balance between having too few opportunities to exercise self-control and too many difficult challenges that cannot be overcome.

An environment with very little opportunity for self-restraint because it is completely lawless or indulgent likely makes for flabby self-control. Permissive parenting situations in which children have no rules to which they must adjust their actions or where every whim of a child is granted may do little to build up self-control in children. In common speech, we call such children "spoiled brats." On the other extreme, parenting in which children's lives and decisions are carefully controlled by parents, teachers, and older siblings may give children too few opportunities to develop self-control because they are rarely given an opportunity to exercise it. Implicitly, such children are being told they are not trustworthy to make their own decisions or exercise self-control. In a sense, children in such situations cannot exercise self-control because everything is

What we seem to need is a balance between having too few opportunities to exercise self-control and too many difficult challenges that cannot be overcome.

other-controlled. Likewise, parental fear that children may fail to exercise self-control may encourage parents to remove so many temptations that the child will almost certainly fail once they actually face them.

Thus far, we are primarily connecting self-control to social and moral behavior. Indeed, the social brain hypothesis sketched in chapter four suggests that all of these deliberate, conscious self-regulatory activities led to the development of large prefrontal cortexes because of pressures from social living. From this perspective, self-control is primarily a tool for social living. Nevertheless, self-control also allows humans to meet the demands of their niches in ways that are not strictly social. For instance, self-control allows one to use planning and stealth to capture prey. Trout ticklers use their human self-control to slowly and quietly enter very cold trout streams and remain still for long periods of time while they gently place their hand under a trout to tickle its underside until a sudden moment of capture. Self-control, particularly as seen in patience, perseverance, and courage, undoubtedly facilitated other adaptive problem-solving in human prehistory. We turn to these virtues next.

RELATED VIRTUES AND EMOTIONAL REGULATION

Kirk is a former US Army special forces soldier.[13] He has faced many harrowing challenges. He has jumped out of planes, snuck into hostile environments, been shot at, and faced numerous other fears with bravery. After his military service, Kirk became a high-ropes course trainer and inspector. He spent much of his days in treetops clipped to cables and dangling dozens of feet above the ground so other people could be coached through obstacles and grow through the process. Surprisingly, Kirk is afraid of heights. Asked why he chose a career that forced him to climb trees, he answered, "It is important to face and manage your fears."

Self-control is related to a number of virtues that thinkers from Aristotle's day onward have considered part of the good life, or thriving. Some of these

---

[13]Kirk's name has been changed to protect his anonymity.

virtues include patience, perseverance, resiliency, and courage. Psychological research shows that perseverance—or "grit"—predicts higher degrees of achievement, life satisfaction, and other positive outcomes.[14] The ability to labor diligently now in hopes of an abundant harvest later has transformed humans from hunter-gatherers to horticulturalists and then settled agriculturalists, and this ability to put foresight and planning into action was made possible by patience and perseverance.

All of these virtues, especially courage, require a heavy dose of emotional regulation, a special type of self-control. Emotional regulation is the ability to monitor, evaluate, and modify one's emotional states to better accomplish one's goals.[15] Like many areas of expertise, what may begin as a conscious effort to control one's behavior or actions can become seemingly automatic with practice. Repeated efforts (and sometimes improved techniques) for regulating fear, for instance, may eventually allow one to become a courageous person not paralyzed by fear. At a certain point, firefighters, police officers, and soldiers almost automatically run toward dangers instead of away from them. Courage allows someone to focus on a goal even when feeling threatened or fearful.

An instinctive fight-or-flight response to threatening situations is a normal part of the animal experience and is experienced as fear or anxiety in humans. In its primitive form, this instinct triggers a physiological reaction that focuses the senses, heightens alertness, and readies muscles for action. Then, depending on the situation, the animal (human or otherwise) is prepared to run, fight, or freeze (and then run or fight if necessary). The extra sugar, oxygen, and adrenaline in the bloodstream will be burned during the fight or flight, allowing the animal to return to normal resting levels before long.

Humans, however, do not have to fight or flee whenever they feel fear. Humans can short-circuit this instinct by consciously reevaluating the situation and deciding on a different course of action. To a degree unseen in other animals, humans deliberately put themselves in fear- or

---

[14]Duckworth & Gross, 2014.
[15]Thompson, 1994.

anxiety-producing situations because they have decided it is better to face the situation than run. Humans restrain themselves from acting aggressively, opting instead to use soothing words, to exact revenge at some later point, or to defer to authorities to resolve conflict. The distinctive human docility and sociality described in chapter three is due in part to emotional regulation. That is, human regulation of emotion facilitated new solutions to age-old fitness problems. One did not have to just fight or flee but could now engage in more deliberate and even socially distributed problem-solving. We do not have to address conflict or danger merely through individual strength but may address these threats collectively by designating roles and procedures for dealing with them, such as through justice systems and militaries. These solutions, in turn, permanently changed human niches.

THE SELF-CONTROL GAP

Self-control, understood broadly, enables us to choose among competing motives, override or tamp down undesirable impulses, or amp up and act on impulses or convictions that are weaker than we would like them to be. Self-control includes deliberately regulating emotions. Self-control is a bit like a muscle that is used in many different domains. Like a muscle it needs exercise; we need practice exercising control over ourselves by making tough choices, overriding impulses, or acting when a big part of us does not want to. But this muscle can be ineffective from underuse and also from being placed in situations in which it is almost doomed to fail.

Our special form of sociality, the acquisition of information, and self-control have together enabled us to build contemporary human niches with all of their wonder and complexity. It may be, however, that for all the good our self-controlled human nature brings about, it sometimes changes our niche such that new challenges for self-control and virtuous decision-making arise. Take time-based punctuality as an example.

As far as we know, since there have been humans, humans have had to exercise self-control to do things "on time." If you were an angler, you might have had to be out fishing before the warmth of the day drove fish to deeper water. If you were gathering, you might have had to get the acorns before the deer and wild turkey got them. As a horticulturalist or farmer, getting seeds in the ground early enough (but not too early) was a key to harvest success. Likewise, being "on time" might have involved getting to shelter before a storm or securing supplies before night-time scavengers were out. But in these sorts of examples, being "on time" is tied to clear cause-and-effect outcomes and related risks and rewards. As the saying goes, "The early bird gets the worm." This kind of punctuality is easily motivated. Many animals do something like it.

At some point in our human past, however, our ancestors discovered the utility of using our ability to be punctual to solve problems for which the timing was more arbitrary. Coordinated activities could be linked to a time of day or phase of the moon for the sake of efficiency. Saying, "Let's meet when the sun is at its highest point the day after the moon is new" can clarify more precisely when it is that a group is meeting up for some activity, thus reducing the waiting time and associated costs. However it came about, punctuality surely played a role in solving large-scale coordination problems and became regarded as a virtue in some human groups. Our self-controlled nature was harnessed to solve niche-related challenges through punctuality. A byproduct, however, is that in the human societies most of us now live in, being "on time" is no longer an extra tool to use when it's helpful but a social or even legal requirement. And the technologies for telling time of day, day of the week, and month of the year are part of a cultural package that must be acquired and used effectively to navigate contemporary societies.

What started as a nifty invention to help solve some problems is now a necessity many of us must obey, almost slavishly. And so children must be taught how to tell time and date, and we all must learn how to force ourselves to be punctual and do things fast enough. We use social pressure to enforce the rules of timeliness. The result? We stress over being late and we rush to do things on time, increasing blood pressure and risk of accidents. How many driving deaths result from someone who was in a hurry to be "on time"? How many heart attacks are provoked by needing to meet an arbitrary deadline? We have niche-constructed a new gap with our nature that we have to learn how to bridge.

My point is not to degrade this niche-constructed virtue but to look at it a new way. In many ways we have used our amazing abilities for self-control to create new environments in which what we *can* do becomes what we *must* do. New cultural gadgets have to be invented and imposed to help everyone else get to that floating platform the kids before us pushed too far out. The result is that we need even stronger self-control mechanisms to thrive, but how do we cultivate such self-control?

A plausible hypothesis based on what we know is that many contemporary societies have created environments that fall outside the optimal balance of opportunities for exercising and effectively using self-control. That is, compared with our ancestors, perhaps even just a few generations ago, many contemporary humans in urbanized, developed nations may be experiencing a vicious mix of too few modest challenges that would build up self-control with too many great challenges that prompt us to succumb.

My thesis is not that some human society once upon a time was the perfect context for the effective and virtuous use of self-control. Rather, a byproduct of the kinds of societies our self-controlled natures have produced is ever-changing and sometimes increasing demands on our effective use of self-control. As self-control has been used to bridge the nature-niche gap, it has sometimes widened that gap by changing the niche.

Our ability to use self-control to regulate our social relationships and build cooperative social networks has enabled the growth of larger societies with a greater diversity of abilities, needs, and perspectives. But this greater size and diversity of societies has created pressure to more extensively codify rules of how people in such societies should get along. And this process of making informal arrangements and expectations more formal and detailed enables (even if it doesn't necessitate) social rules and laws that are less and less intuitive. That is, the distance between morally relevant intuitions and the rules becomes greater, requiring more conscious attention and self-control to understand and abide by them. Furthermore, the greater the level of societally imposed regulation, or political control, the less individuals need to reason through what to do and how to behave for themselves. They don't have the same practice in exercising self-control because they are socially or politically controlled.

Temptations to act in ways regarded as moral violations were probably plentiful in ancestral human societies, but several factors may have helped place the bulk of temptations into the range that people could readily grow to handle. First, decision-making, including moral decision-making, may have been closer to intuitive foundations, thereby not overtaxing conscious attention and self-control. It is easier to submit to the authority of a known and esteemed leader telling you that leaving human waste in a particular part of the forest will scare off game than an anonymously written sign telling you not to park on one particular section of one side of the road on the first and third Thursday of the month between 6:00 a.m. and 1:00 p.m. unless it is a state-recognized holiday. It is easier to feel compelled to share food in a roughly equitable fashion with the other people in one's village with whom you interact on a daily basis than to feel a moral compulsion not to buy those particular shoes because they have allegedly been manufactured by anonymous people in another part of the world who are not treated or paid well. We have niche-constructed societies with increasingly plentiful rules that are causally and morally opaque.[16]

A second factor of many contemporary societies that may be frustrating our abilities to build up and deploy sufficient self-control is that the rules for behavior are constantly changing. Rapid growth of communities in terms of size and diversity creates political pressure to address specific points of social tension and, hence, create new laws and regulations.[17] Add that people are much more transient, mobile, and likely to

[16]The popular television series *The Good Place* took playful aim at this dynamic when its main characters discovered that no one had been good enough to make it into the good place—even the one person who figured out the points system—for hundreds of years. Why? In season three we learn that society has become so complex that even the simple act of buying a tomato can be a morally negative action.

[17]For those who doubt that contemporary societies have more complex systems of rules governing human behavior, consider that the governor of California approved 870 new laws in 2019 alone

change jobs than in the past.[18] As a result, we cannot simply rely on accumulated knowledge and habits about how to behave. We constantly have to learn what is acceptable here and now.[19]

Likewise, in small-scale societies of the past, lower degrees of anonymity and expectation of successful secret-keeping along with a more uniform distribution of goods likely led to far less self-control defeat than is commonly experienced in today's urbanized societies.[20] In smaller-scaled ancestral living, reasons to resist temptation were a natural extension of everyone knowing everyone else and it being very difficult to keep secrets. The cousin's beautiful cape may have been tempting, but if one did steal it, then everyone in the village would know about the theft. The would-be thief could never wear it. The neighbor's spouse may have been tempting, but the risk of being caught in a tryst by a friend or family member of the absent neighbor would have been high enough to discourage frequent carelessness. Taking more than one's fair share, or not giving enough, would quickly be noticed by others, leading to the ruin of one's reputation. In many societies today, it may be much easier to get away with transgressions, so the temptation to engage in them may be greater—often too great.[21]

Too many social regulations may present too many opportunities to fail to exercise self-control and lead to demoralization and giving up (or just cutting corners). High degrees of anonymity and the ability to keep secrets may discourage self-denial. High inequality in the distribution of

---

(Wiley, 2019). A typical Californian has local, county, state, and federal laws and regulations to navigate—many involving reasons that are unclear to the average person and only a distant mapping onto any intuitive moral foundations—as well as general social "rules."

[18]For instance, consider Esipova, Pugliese, and Ray (2013, May 15).

[19]For instance, in his fascinating treatise on Chinese World of Warcraft players, anthropologist Ryan Hornbeck (2013) notes the frustration many Chinese young adults experience over the enormous number of frequently changing social regulations they must navigate in contemporary urban China.

[20]Of course, not all small-scaled societies are or were strictly egalitarian and certainly not free of attempts to keep secrets or deceive. The hypothesis offered here is that as challenging as it was to successfully monitor social expectations and successfully exercise self-control in the past, the demands are likely even greater now.

[21]It may be that high population densities make getting away with some offenses harder, not easier, so the relationship between the size of a society and the anonymity of its members is not a simple one.

goods may promote envy and the temptation to take from others, perhaps even rationalizing antisocial behavior in the name of "fairness."

A fourth factor working against the effective deployment of self-control to live virtuously is that some segments of society use powerful tools to actively defeat our self-control. The advertising industry places heaps of temptation on individuals. Thanks to billboards and digital technologies, urbanites are bombarded by a constant barrage of images, sounds, and smells engineered to part them from their money or succumb to temptation to do things they might not actively look to do if left to their own devices. One might eat less if those alluring smells were not constantly wafting. If so many goods were not attractively arranged in shop windows just within reach, they would be easier to live without. The omnipresent internet only intensifies the pressure. Resisting this temptation all day every day may be too much for many of us to handle.

Given these four challenges for developing and using self-control, consider a day in the life of a contemporary school-aged child in an urban setting. The child may be awakened by an alarm before she wants to rise (instead of automatically and habitually waking with the dawn), get ready for school by eating food she may not want to eat (instead of there really being no choices) and wear clothes she would not necessarily choose because they would be deemed inappropriate (instead of wearing the only set of clothing available), walk to school by navigating many street crossings according to the rules that govern pedestrian behavior (instead of going the way that is most direct), report to the appropriate class at the appropriate time (instead of time being largely fluid), sit in the designated seat, restrain herself from speaking or moving too much for big chunks of the day (instead of moving and talking most of the day), play on the playground according to the rules (instead of inventing loosely structured, improvised games), queue for lunch in the right way, and so forth. Today's children experience many opportunities to exercise their self-control "muscle," but they may also find too many opportunities to fail as well, thereby undermining their confidence and motivation to build up the self-control muscle.

Similarly, many adults spend all day or all week in heavily regulated jobs doing things they lack little intrinsic motivation to do because they are so distant from basic survival needs and motivators. They force themselves to keep performing, act as if they care, please their bosses and coworkers, and so forth. They may have to attentively commute over dangerous road systems and otherwise manage their lives using societally prescribed rules and regulations. By the end of the day or week, many adults may have expended a tremendous amount of energy controlling themselves and subsequently may be demotivated to keep trying. The point is not that Stone Age people had more free time on their hands (though they likely did)[22] or that they necessarily enjoyed all the tasks they had to do but that in smaller-scaled societies, there was less need for each aspect of human life to be a series of decisions requiring self-control.

Many societies that have been built at least partly because of humans' extraordinary ability to self-control are, in fact, societies that constantly strain their citizens' self-control. Having little self-control buffer means many people constantly live on the edge of moral failure. No one has enough self-control to go through life always doing what is right and never doing wrong. So how does one live a good, rich, morally satisfying—thriving—life? Leading self-control researcher Roy Baumeister suggests that, in addition to building up self-control muscles, one should use one's limited strength to steer oneself away from temptation before it strikes.[23] Use some self-control and higher-order reasoning to anticipate situations that may present too great a temptation, and make a plan to avoid them. For instance, if the doughnut shop you pass on the way to work is going to be more tempting at the end of a hard day, plan an alternate route home.

If a tempting situation cannot be avoided, exercise some self-control beforehand in deciding how one will act in that situation. To illustrate,

[22]Such a possibility is drawn from ethnographic studies of hunter-gatherers (see, for instance, Sahlins, 1972). The "standard" workweek of forty hours plus time spent on food preparation and various unpaid chores is likely not standard at all among humans past or present.

[23]Baumeister, 2012.

someone trying to reduce his or her alcohol consumption might plan a non-alcoholic drink order before going to the restaurant so that when the server asks, "Can I get you a drink while you look over the menu?" the response, "I'd like a lemonade," is queued up and ready. Once one is in a tempting situation, trying to figure out what to do may lead one to succumb. Implementing a virtuous action already planned before temptation strikes is easier than devising and implementing the same action under temptation.

Another judicious use of self-control resources is to apply them to the development of behavioral strategies and habits so you do not have to fight the same self-control battles over and over again. For example, some might swear off keeping ice cream in the home (for instance, to avoid gluttony) so they will not have to face the ice cream temptation on a daily basis and can devote their self-control muscle to other heavy lifts. Others might awaken every morning at the same time, whether on weekdays or weekends, to make waking up habitual so that conscious self-control resources are fresh and ready to fight other battles.

THE EMOTIONAL REGULATION GAP

Self-control, including emotional regulation, has helped humans solve numerous fitness problems, but those solutions have changed the niches into which humans are now born. Self-control may currently be a more precious commodity and yet also be more threatened by distraction, temptation, and exhaustion than ever before. Emotional regulation in particular may also face a nature-niche gap in many contemporary settings, particularly in relation to regulating fear, anger, and anxiety.

Self-control may currently be a more precious commodity and yet also be more threatened by distraction, temptation, and exhaustion than ever before.

As discussed earlier, fear motivates action. That action in turn helps the physiological rush associated with fear dissipate, allowing the body to return to a normal state. What happens, then, when one is not allowed to act, such as when fighting or fleeing in the face of a threat is not socially acceptable? Self-control helps prevent a child from punching another child when he or she feels threatened, but then the extra adrenaline, blood sugar, and blood oxygen have no simple way to "burn off." An adolescent facing a threatening bullying situation at school every day may have a flight response but no place to run—he or she is forced to stay in the threatening situation. This can result in internalized anxiety and feeling overwhelmed.

Compounding the problem, whereas the proper trigger of the fight-or-flight mechanism was a specific identifiable threat such as an enemy, a large beast, or a storm, the actual domain of threats that stimulate it today can be unseen others or abstract entities, such as government agencies or companies. In our hunter-gatherer or small-scale-farmer past, one may have lived with the looming threat of crops being destroyed by pests or a bad hunt, but one could not be fired, evicted, audited, or bombed by unknown and unseen enemies from thirty thousand feet. People today live with a whole host of looming threats that cannot be evaded or fought directly. Some of these are even illusory threats ginned up by advertisers or politicians and peddled by mass media. As a result, the natural emotional regulation tools available to humans—by virtue of their being humans—are often insufficient for thriving in today's world.

THE DECLINING ROLE OF RELIGIONS?

Current research in psychological, cognitive, and evolutionary studies of religions has begun highlighting the positive role religious practices and participation may play in moral decision-making, altruism, and self-control. For instance, psychologist Jesse Bering has argued that the hypersociality of our species has made us naturally receptive to the idea that unseen beings could be watching and judging us. As a result, we are more careful about how we behave, making us better community members.[24] Similarly, psychologist Ara Norenzayan has argued that belief in powerful gods who police human moral behavior motivated past (and present) peoples to treat each other well and thereby facilitated the building of large-scale societies.[25] Anthropologist Richard Sosis, among others, has produced evidence that engagement in public religious practices may encourage greater levels of in-group trust, cooperation, and altruism. This ability of religious behaviors to improve fitness in groups may be part of how religious systems evolved.[26]

A broad analysis of existing literature also suggests that high levels of religious participation correlate with higher levels of self-control, another way in which religious practice appears to be adaptive.[27] It may be that conventional religious practices, such as developing the habit of getting dressed and out of the house for services, engaging in prescribed ritualized behaviors, paying attention to sometimes boring readings and teachings, and adopting other religious disciplines such as daily prayer and Scripture reading, are collective self-control exercises that nonreligious participants miss out on if they don't create substitutes that generate enough motivation to do the job. In sum, many current scholars suggest that religion may be a prominent part of the human experience because of its general adaptiveness for humans living in groups. Humans who were religious may have had greater fitness than humans who were not.

Humans have likely been engaging in thought and behavior that we commonly call "religious" for anywhere from thirty thousand to a hundred thousand years or more.[28] We have wondered about transcendent realities, gods, and afterlives for much or all of that time. Religious beliefs and practices have been a cultural resource for helping people regulate themselves and behave in ways beneficial to themselves and their

---

[24]Bering, 2011.
[25]Norenzayan, 2013.
[26]Sosis & Ruffle, 2003.
[27]McCullough & Willoughby, 2009.
[28]Henshilwood & D'Errico, 2011; van Huyssteen, 2006.

groups.[29] The human niche has been religious. What happens, then, when this religious niche is dramatically changed in some cultures? The demise of religious life in the world is often greatly overstated; nevertheless, the rate of active religious participation in many Western nations has seen a decline in the past fifty years.[30] As a result, fewer people may be exposed to the self-control–shaping influences of religious practices and teachings than in past decades or even in the history of humanity. It may be that other nonreligious sources of morally relevant self-control practices could step into the space left by religion's retreat. If, however, humans have naturally received self-control practice and moral encouragement through collective religion, the change in niche may leave a gap. Many of us take vitamin supplements to try to make up for a deficient diet, but many of these supplements are not as readily absorbed as natural foods because of their mismatch with our digestive systems. Analogously, will secular solutions carry reduced benefits due to an ill fit with human religious nature? It is not remotely clear that secular institutions can play the same role in encouraging the development of self-control and moral behavior that many religious traditions have sought to cultivate.

FINDING THE SELF-CONTROL GOLDILOCKS ZONE

In addition to being a remarkably social species that avidly acquires information and expertise to solve problems, humans also display considerable self-control. We humans consciously try to wrangle our emotions, our impulses, and our thoughts to act more tactically and, sometimes, morally. Fortunately, when used strategically, our self-control capacity can be increased through successful exercise, but it may also be discouraged through failure or feelings of helplessness and overwhelmed in the face of too many challenges. We can even experience debilitating and uniquely human emotions such as shame. Like Goldilocks finding

[29]Of course, there are always exceptions and sometimes horrific ones, but we are talking about general patterns.

[30]See, for instance, Pew Forum on Religion and Public Life (2008, February) and Pew Forum on Religion and Public Life (2012, October 9).

the bed that is not too hard or too soft, or the porridge that is not too hot or too cold, it is not easy for us to find the optimum between enough opportunity for growth through challenges and a high enough rate of success because the challenges are not too numerous. Using our distinctively human self-control abilities to find this sweet spot for our children and ourselves may be a critical key to thriving.

7

A SUMMARY AND A PUZZLE

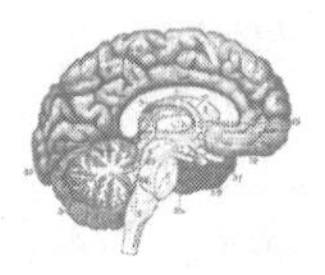

To recap the argument thus far, thriving is approaching what one is meant to be, moving ever closer to one's purpose or telos. We thrive more the more we move along the path to becoming the fullest version of ourselves, but being that full version of ourselves may be in the distant future, perhaps not even in this life.[1] Thriving, then, has a right-now and a not-yet dimension to it.

Furthermore, I have argued that we get more details concerning human telos and how to pursue it (and, hence, human thriving) by considering what human nature is, particularly in relation to human environments. A living thing cannot be expected to have a telos that falls outside its nature. Specifying aspects of our nature helps identify the kinds of things we, as humans, may be reasonably able to do, and evolutionary psychology is a useful tool in discovering fresh insights into what makes humans human. For instance, the fact that we are such deeply social animals leads to the observation that human thriving has both an individual/personal dimension and a collective/social dimension. It is very difficult to thrive if our group or community is not thriving, and our community is not thriving if the individuals in it are not thriving.

---

[1]King, in press; King & Defoy, 2020; King & Whitney, 2015.

It is tempting to think of human nature as an isolated bundle of features or capacities, but an evolutionary perspective encourages us to consider not just typical traits, but also how they fit—or don't—with a particular environment or niche. An obstacle to thriving may be a bad nature-niche fit: when the tools one possesses to solve basic fitness problems are inadequate, it is hard to thrive. So thriving involves an ever-better fulfillment of our nature in addressing challenges posed by our niches.

This analysis of thriving would apply up to this point to many or most nonhuman living things. A common gray squirrel, for instance, would be thriving if it was behaving ever-squirrellier. We could discern what it means to be squirrelly, in part, by considering the physical and behavioral traits of a squirrel from an evolutionary perspective. This scientific perspective might reveal that the nature of a gray squirrel is most actualized and satisfied when the squirrel is in an environment with enough oaks, piñon pines, and other trees to provide it food and shelter. A healthy, very squirrelly gray squirrel in the usual sense, however, would not seem to be thriving if it were suddenly in a treeless urban environment. Its nature and niche would have a mismatch or gap.

Humans are different from squirrels, however, not just in how we might characterize human nature but also in when and how we face a burdensome nature-niche gap. The difference is that humans are radical niche constructors, intentionally or not. That is, we humans can change our environments so dramatically in a single generation that the demands on our children may be very different from what we faced growing up. For most other species in natural history, the subtle changes one generation has had on the environment of subsequent ones have been accommodated by migration, minor amounts of very targeted social learning (in a small number of species), or the species being shaped by natural selection to manage the new niche conditions. Not so with us humans. Not only do we modify our behaviors for our niches, we modify our niches to meet our desires. In fact, our power to shape and reshape our environment is so typically dramatic that it may be considered an ordinary product of our nature.

What, then, are the features that characterize human nature? I have grouped them into three interacting and overlapping bundles or capacities: (hyper)sociality, extreme gathering of information leading to individual and local expertise, and conscious, deliberate self-control. These three clusters make up the characteristic ways we attempt to bridge the nature-niche gap, but they also enable our tendency to radically shape our niches. They are both the source and the solution to the perpetual nature-niche gap. In our attempt to improve our lives in our environments we continually present ourselves with new challenges—chronically increasing our social, expertise, and self-control demands. We can't seem to help ourselves—we continually fiddle with our niche.

In our attempt to improve our lives in our environments we continually present ourselves with new challenges.

This recap ends with a puzzle. If thriving is in part living into our nature and a large nature-niche gap is an obstacle to thriving, but our nature seems to spawn just such a gap, aren't we left with a paradox or at least an insurmountable challenge? It seems like being human—actualizing our human nature—unswervingly creates a large and ongoing nature-niche gap that in turn gets in the way of human thriving. Does this mean we humans are pitiable creatures who can never thrive?

WHAT EVOLUTIONARY PSYCHOLOGY DOES NOT TELL US

A common ethical mantra is "ought implies can"; if it can be rightly said that you ought to do something, then it typically has to be something that you are able to do, at least in principle. For example, it would be strange to say someone ought to fly into a burning building to save a baby—that he or she has some kind of moral or normative obligation to do so—if the person in question cannot fly. To make such a claim would be unreasonable. "Ought," then, implies "can." In contrast, the fact that something is the case does not imply it ought to be the case. Just because someone tends to get in a sour mood and mistreat others when hungry does not mean they should. "Is" does not imply "ought." Similarly, just because something is the case does not necessarily mean that something must be the case. So far in our discussion I have stayed fairly close to the facts evolutionary psychology presents as our best understanding of what human nature is: what capacities humans typically have, how humans typically and distinctively behave, and the like. One of these facts is that humans radically impact their environmental niches. However, just because this is what we humans do does not imply that it is what we ought to do. We will inevitably niche-construct, but our impact on our niches need not be so dramatic.

A careful consideration of evolutionary psychology helps us better understand the parameters within which humans are able to discover their telos, pursue it, and thrive, but evolutionary psychology is a descriptive scientific enterprise and alone cannot tell us what we should, ought, or must do. Evolutionary psychology can help us better understand our nature, but it cannot tell us what to do with it. In fact, even though it has a lot to say about how our ancestors have gone about surviving, reproducing, and rearing children, evolutionary psychology alone cannot even tell us that we *should* survive, reproduce, or rear children! Evolutionary psychology describes for us only what we do when we do it and what the consequences might be on our fitness. Evolutionary psychology is a powerful tool, but like all tools, it is powerful for what it was designed to do and not necessarily anything else. A telephone is a great tool for person-to-person communication, but it is a lousy tool for telling you what you should say to another person. Similarly, evolutionary psychology is a great tool for better understanding what capacities we humans have that make us unlike other animals, but it is a lousy tool for telling you what you should do with that information. What is needed is a normative framework that guides behavior or establishes an ethical standard or norm. Such a framework or ideology serves as a prescriptive, not merely descriptive, intellectual tool to take us the rest of the way in our treatment of thriving. For this normative framework, I turn to Christian theological anthropology.

THEOLOGICAL ANTHROPOLOGY

Theological anthropology is the area of theology that concerns both what humans are and what they should be from a theological perspective—it is what we understand God to be telling us about what humans are and should be.[2] At its best, the field draws insight from the various ways God communicates truth to people. These ways include specially inspired Scripture (the Bible), our own experiences interacting with God and the experiences of others, careful observations of the world around us and reasoned reflection on it (including the sciences, philosophy, and other scholarly areas), and the cumulative insights of those who have come before us and attempted to consider all these sources of God's communication to us (i.e., tradition).[3] Theological anthropology, then, has a place for sciences—even evolutionary psychology—but does not depend on the sciences alone. A review of theological anthropology would move us far beyond our purposes in this book, but a quick tour of some neighborhoods in this area of theology will be useful in addressing the apparent human nature-niche gap problem.

One old but still lively area of theological anthropology concerns what it means to be human and particularly what the Bible means when it says that humans were created in God's image (*imago Dei*). Genesis 1:26-27 says:

> Then God said, "Let us make man in our image, after our likeness; and let them have dominion over the fish of the sea, and over the birds of the air, and over the cattle, and over all the earth, and over every creeping thing that creeps upon the earth." So God created man in his own image, in the image of God he created him; male and female he created them.

How should we understand what it means that "in the image of God he created him; male and female he created them"?

---

[2]King, 2020; King & Defoy, 2020; King & Whitney, 2015.

[3]These four ways (Scripture, experience, reason, and tradition) are known as the "Wesleyan quadrilateral." But these aren't four independent, non-interacting sources of revelation or insight. For instance, we read Scripture with interpretive assumptions of a particular tradition, using reason, and shaped by our personal experiences.

Commonly, the question about what it means for humans to be created in God's image is anchored by two assumptions: first, each and every human is a bearer of God's image (or all humans are *imago Dei*), whether or not they are suffering from a developmental disorder or a debilitating disease or injury, and, second, no other living things that have existed on earth since the writing of the book of Genesis are *imago Dei*. This second assumption comes from how the Genesis account appears to contrast humans with the other animals: human creation is described after the other animals, demarcated as "in our image" and then human "dominion over" the other animals is described in the same breath.

It does not follow that human uniqueness and being *imago Dei* are one and the same thing: humans may be unique in their ability to play video games but that ability is not what leads us to be declared to be created in God's image. On the other hand, it could be that, prehistorically, another species shared many of the same features that now seem to be unique to humans. Perhaps these beings, too, were also *imago Dei* but were extinct by the writing of Genesis and so irrelevant for the purposes of the biblical text or any oral tradition that immediately predated it.

If all humans but no current nonhumans are *imago Dei*, what exactly is it that makes us so? This question has been the concern of theologians for centuries and so I do not propose a comprehensive treatment here. But part of an answer comes from affirming that Christ is the perfect image of God, an idea synthesized from various New Testament passages (see Colossians 1:15; 2:9; Hebrews 1:3; John 10:30). Pam, my collaborator, and William Whitney contend that a telos shared by all humans is being conformed to the image of God in Christ.[4] They suggest that the life of Jesus Christ offers a direction toward which humans grow and mature in the Spirit. Patterning our life after the character and way of Jesus' life offers a way to become more like the perfect image of God.

---

[4]King & Whitney, 2015. Note, too, that observing that Jesus is the perfect image of God that each of us should imitate pushes away from the idea that somehow humans only collectively, but not individually, image God. Jesus of Nazareth was an individual, after all (see also Balswick, King, & Reimer, 2016).

As Christians, we are called to live a life in the Spirit according to the pattern of Jesus, but this does not answer our question, what makes humans *imago Dei* and not other species? The sorts of answers that have been offered are instructive, however, and can be divided into two basic sorts: properties and functions. A property approach to the question tries to identify just which property or properties humans have that make them *imago Dei* or, more strongly, what property is the image and likeness of God in us. The functional approach tries to identify the roles, responsibilities, or functions of humans that represent or imitate the actions or purposes of God.

Patterning our life after the character and way of Jesus' life offers a way to become more like the perfect image of God.

Property approaches to the image of God. The properties that have been suggested as critical to humans being in the image of God may be grouped into four clusters.[5] First, humans are regarded as being like God in that they are rational, intelligent, or capable of exercising wisdom. In fact, our species name, *Homo sapiens,* means "wise human." Second, human freedom has been thought to be the way in which we image God. God is perfectly free (one definition of being omnipotent or all-powerful), and humans reflect this freedom in being free in some important ways we do not see in other species, able to control themselves and choose to act or not act in specific ways. Could it be that we are similar to God but different from other animals in that we can freely and willfully choose our actions? Third, being language users, a social and cultural trait, has been pointed to as a way in which humans may image God. After all, God speaks the cosmos into existence, Adam goes about naming the animals, and we read that Adam and Eve talked with God. A fourth property or capacity is the ability to form and maintain interpersonal relationships and love others. This fourth property is probably the most commonly emphasized by contemporary theologians, as we will discuss below.

[5]For a review and similar analysis, see Thiselton (2015), Barrett and Jarvinen (2015), and King and Whitney (2015).

Interestingly, these properties of humans that have attracted the attention of theologians studying human nature show some points of convergence with our analysis of human nature from the perspective of evolutionary psychology. The temptation to identify the image of God with human rationality, wisdom, or linguistic capabilities is understandable given the extraordinary capacity humans have to engage in reflection and form metarepresentations (described in chapter four) and our unusually developed capacity for complex language use (as reviewed in chapter three). Similarly, it seems sensible to highlight human freedom and the morality that freedom enables as distinctive, given human moral intuitions and our powerful potential for using our remarkable self-control to build on those intuitions and make moral decisions before acting (see chapter six). What's more, the unusually social character of humans, including our ability to love others selflessly, has not escaped the notice of evolutionary psychologists either (chapter four).

These points of resonance suggest that our evolutionary analysis is on the right track, but still, identifying distinctively human properties does not tell us what to do with those properties. They still do not provide a telos or purpose. The functional approach, however, to the *imago Dei* provides more direction.

Functional approaches to the image of God. Theologian Anthony Thiselton observes that though varying degrees of the properties identified in the previous section effectively separate humans from nonhumans, it is imprudent to isolate distinct properties or capacities as sufficiently and completely capturing the *imago Dei.*[6] He argues that the Bible points to no specific human capacities as characteristic of the image of God; rather, imaging God occurs in our wholeness, not in our parts. One reason to reject the idea that some magic part of being a human is also the seat of the image of God is that once we consider infants, those with developmental disorders or brain injuries, and those suffering from

---

[6]Thiselton, 2015.

Alzheimer's disease, it is hard to find any feature that each and every human has that another animal does not have.

Aware of the dangers and limitations of identifying one feature or capacity that is somehow *imago Dei,* many contemporary theologians have begun emphasizing functions, roles, statuses, or purposes God may have bestowed on humanity. Two of these functions are most commonly advocated: humans as chosen for a special loving relationship with God and humans as God's representatives in "ruling" over the rest of creation.

Perhaps because of the unique bundle of properties that we humans commonly exhibit, including our ability to form personal relationships, God chose to love us personally. We functionally image God in being granted the ability to love God and love others in a special sort of way. That is, when God says, "Let us make humankind in our own image," God was selecting a particular group of living things—humans—to have a special relationship with Him and, consequently, with the rest of creation. Though God may have selected humans because their particular typical bundle of capacities made them capable of being image bearers, it is God's choice to form a relationship with them and imbue them with some extra status that makes them the image of God, or able to image God. Nevertheless, in this view, God's decision does not rest on what features each and every human may or may not exhibit. God can choose to love us all and each in a special way regardless of our stage of development, disability, or injury.

This approach to the image of God is a relational one but it differs from merely observing that humans have the property of being supersocial or relational. Humans are not in God's image because they are especially social. Rather, because they are appropriately social (perhaps among other characteristics), God chooses humans from among other creatures to have a special social relationship with. Many have supported this relational emphasis by reference to the trinitarian character of God, that God is three persons in one. That God is quoted in Genesis as using the first-person plural pronouns *us* and *our* is regarded as textual evidence for such a view. For instance, theologian Stanley Grenz posits that "the three trinitarian persons are persons-in-relation and gain their personal identity by means of their interrelationality," thus highlighting fundamental human essence as persons-in-relation.[7] Similarly, with Jack Balswick and Kevin Reimer, Pam has followed theologians such as Karl Barth and Miroslav Volf in regarding *imago Dei* as having a relational core. As the trinitarian God necessarily combines both uniqueness of persons and unity of the Godhead, so too humans are created to be unique individuals united with God and with others, akin to the I-Thou relationships championed by Barth and Jewish philosopher Martin Buber.[8] Balswick, King, and Reimer describe human telos as reciprocating selves, writing:

> To live as beings made in the image of God is to exist as reciprocating selves—as unique individuals living in relationships with others . . . as a distinct human being in communion with God and others in mutually giving and receiving relationships.[9]

Thiselton sees promise in such a relational approach and maintains that this relationality is initiated through the loving extension of God, allowing communion with God and, subsequently, communion with others in love. The moon has the right properties to reflect the sun's light to the earth, but it can do so only if the sun shines on it.

A second common functional approach to the question of what it means for humans to be created in God's image comes from a fairly straightforward reading of the Genesis text. After God says, "Let us make humankind in our image, according to our likeness," the very next statement is "and let them have dominion over the fish of the sea, and over the birds of the air, and over the cattle, and over all the earth, and over every creeping thing that creeps upon the earth" (Genesis 1:26-27 NRSV). Psalm 8 echoes this theme that the creation of humans just beneath God is associated with rule over the other animals:

---

[7]Grenz, 2001, p. 9.
[8]Balswick, King, & Reimer, 2016; King, 2016.
[9]Balswick, King, & Reimer, 2016, p. 36.

What are human beings that you are mindful of them,
mortals that you care for them?
Yet you have made them a little lower than God,
and crowned them with glory and honor.
You have given them dominion over the works of your hands;
you have put all things under their feet,
all sheep and oxen,
and also the beasts of the field,
the birds of the air, and the fish of the sea,
whatever passes along the paths of the sea. Psalm 8:4-8 (NRSV)

Concerning the *imago Dei,* theologian Herman Bavinck writes, "What this [the *imago Dei*] means is not fully stated, but it does include human dominion over all of the created world in conformity to God's will."[10] This particular functional interpretation of the image of God must be considered with care because of the autocratic connotations that "ruling" or "dominion" may carry in contemporary ears. Balswick, King, and Reimer extend their understanding of reciprocating selves to becoming humans who live in appropriate reciprocity with God's creation.[11] Pam and I align with Richard Mouw, who contends that imaging God involves right acting.[12] Drawing on Kuitert's description of humans as "covenant partners," Mouw notes that bearing God's image involves living in covenant, pointing to the importance of relationship while in partnership. It points to the privilege and responsibility of living with God, humankind, and creation. Human rule or dominion is to reflect God's rule and kingship and it should represent genuine care for the created things that God has declared "good" in Genesis 1 and not abuse and exploitation. Human "dominion" should be exercised in accordance with God's wishes and not for merely human ends. Human thriving should be characterized by peace or shalom with their social and natural environments.[13]

---

[10]Bavinck, 2011, p. 284.
[11]Balswick, King, & Reimer, 2016.
[12]Mouw, 2012.
[13]For a similar notion, see Volf and Croasmun (2019). Michael Burdett (2020) persuasively argues that a functional view of *imago Dei* that focuses on the human role as God's stewards is attractive

More specifically than having dominion, others have also suggested that a functional interpretation of the *imago* involves vocation. Vocation may be understood as the role or manner by which specific individuals or entities fulfill God's command to have dominion over the earth. We cannot all be intended to care for creation in the same manner. Synthesizing relational and functional views described above, Pam points to the importance of human uniqueness and relatedness in understanding vocation.[14] Drawing on the telos of the reciprocating self, vocation involves becoming more like Christ—the perfect image as our unique selves as we relate and tend to the local and global world. Thus, vocation lies at the intersection of one's unique skills, capacities, sparks, and passions; the needs of human others and the broader world; and the ways one is being conformed to the image of God in Christ. Our vocation is the way we engage in God's ongoing work of creation, redemption, and bringing to completion His aims for this world.[15] As we image God, we should do so in a manner suited to our unique natures, that is, meaningfully relating to and contributing to the greater world and enabling us to become more Christlike. Our specific vocations are the means by which we participate in God's kingdom. This view of imaging God emphasizes being, relating, and doing as part of human telos.

As strong as the textual evidence may be for being God's representatives or stewards in caring for the rest of the created world, such an interpretation has the difficulty that it seems to require certain properties or one cannot fulfill the function. For instance, regarding the human capacity to exercise dominion, philosopher Nicholas Wolterstorff writes:

> A good many human beings do not have the capacities necessary for exercising dominion. Those who are severely impaired mentally from birth never had them. Alzheimer's patients no longer possess them. Such human beings neither resemble God with respect to possessing those

---

because of its relatively light amount of friction with human evolution and also its harmony with recent scientific literature emphasizing humans as hyper-niche constructors.

[14]King, in press.

[15]King, in press. Similarly see Volf and Croasmun (2019).

> capacities nor can they implement the divine mandate or blessing by employing those capacities.[16]

Though being in a loving relationship with God is not as sharply subject to this problem—God could choose to specially love someone who is profoundly impaired—the fact is that if some individuals lack the capacity to reciprocate God's love in some form, then it is hard to see how they are exercising the function that has been proposed as the core of being in the image of God. It seems, then, that functional approaches to the *imago Dei* may be vulnerable to similar criticisms as property approaches: just as not everyone at all times manifests requisite properties, every human at all times doesn't manifest the ability to serve the same functions.

To solve this sort of problem, Wolterstorff appeals to an underlying human nature and not necessarily properties that are manifest in each and every human at each and every point in our lives. That is, there is something like a grand design for what a human should be in the fully realized form, a design not necessarily present at all stages of development or in all humans. A bit like a builder's blueprint for a house that can be made manifest in a variety of ways, all human beings have the same human nature in spite of our differences and whether it is malformed or in process of development. It is because of this nature or blueprint that God can say humans are created in the image of God and have intrinsic worth. After all, none us lives up to the full potential of our nature this side of eternity, so it is not our performance or the fulfillment of our nature that gives us worth but our God-given telos. It is this nature that, when functioning properly, enables us to form an appropriate loving relationship with God and others and represent God's caring rule over the natural world, perhaps because of our God-endowed properties of rationality, language, free will and morality, and sociality. A perspective on the *imago Dei* that emphasizes a God-given human nature or telos can capture the strengths of both property and functional approaches.

---

[16]Wolterstorff, 2008, p. 249.

To sum, theological accounts of what it means to be human typically concern how to understand being created in the image of God, and this image has been understood in many different ways. Human properties that have been suggested as critical include our rationality, use of language, morality and free will, and sociality. Even if we are correct in identifying properties that set humans apart as capable to be designated by God as image bearers, merely identifying these properties does not tell us what they should be used for. The two primary functions that have featured prominently in discussions of the image of God are (1) being in an interpersonal loving relationship with God and others and (2) serving as God's representatives or stewards in caring for his creation. These two functions are to be lived out through our specific vocation. The directional interpretation of the *imago Dei*, which emphasizes becoming Christlike, coupled with these functional perspectives on the image of God gives us a more complete direction, purpose, or telos for our distinctively human bundle of capacities: to love God, love others, and care for God's world as his representatives individually and collectively.[17]

As with the distinctive properties of humans, these two functions (loving God and others and living out vocation in caring for the creation) have some resonance with our evolutionary analysis. The unusually social character of humans, including our ability to love individuals, has already been noted. This social character, combined with our intuitive moral foundations (most of which pertain to interpersonal interaction) and our natural cognitive systems for attachment and forming intimate bonds through social grooming, provides us humans with an amazing toolkit suitable for loving each other personally as precious individuals. As previously noted, having such loving relationships predicts other indicators of well-being.

These facts suggest that proper use of our distinctively human properties to form a deep and broad network of loving relationships is one aspect of what humans should do, a point reaffirmed throughout Scripture

[17]King, 2018; King & Whitney, 2015.

and Christian theology. For instance, Jesus taught to "love your neighbor as yourself" (Matthew 22:39) and said, "Therefore, you should treat people in the same way that you want people to treat you; this is the Law and the Prophets" (Matthew 7:12 CEB). Understanding in more scientific detail what properties humans have for forming these loving relationships may give us better insights in how to love better, especially under difficult circumstances, but it is Christian theology that tells us that we should do so.

Through a similar analysis of the emerging sciences of cultural evolution and evolutionary psychology, philosopher Michael Murray persuasively arrives at a similar conclusion. In his essay "Reverse Engineering the *Imago Dei*," Murray writes:

> So what does this show? I claim that it shows that our evolutionary history seems to have been set up in a way that pushed us down an evolutionary path that allowed for the emergence of creatures with the critical mental abilities needed to be able to engage in relationships of love and friendship. We might thus see this evolutionary history as one that was crafted for the purpose of yielding creatures that are made for the very purposes God intended for us, and perhaps for manifesting the divine image.[18]

Humans are unusual among existing creatures in having certain properties that make them capable of answering God's call to love each other and to love God, in part by serving as His stewards of the created world.

***Relating to God and the* sensus divinitatis.** These same natural properties for loving other humans are also the equipment for relating to God, or at least so say numerous scholars from the cognitive and evolutionary sciences of religion, including those who are not theists themselves.[19] Many scholars working in this area have concluded that religious beliefs of various sorts (e.g., that there is at least one superhuman being, that we can interact with these beings through rituals, that some aspect of humans might continue after death) are largely a natural product of how

[18]Murray, 2019, p. 18.
[19]For example, see Bering (2011), Barrett (2012), and McCauley (2011).

human minds work in ordinary human environments.[20] Part of this ordinary mental work is to regard features of the natural world as having purposes, purposes most explicable in terms of someone(s) designing or creating them.[21] Some scholars working from a decidedly evolutionary perspective have even suggested that these religious beliefs and practices, once they appeared in human groups, offered considerable fitness advantages to those who had them.[22] Religious beliefs and practices, then, may have played a critical role in the evolution of human social groups.

Humans are unusual among existing creatures in having certain properties that make them capable of answering God's call to love each other and to love God, in part by serving as His stewards of the created world.

The alleged "naturalness" of religious thought or even belief in God has led some scholars to suggest that scientists are now generating independent evidence for what theologians in the Reformed tradition have called the *sensus divinitatis* or "sense of the divine."[23] The idea that some kind of vague, imperfect sense of divinity is a product of ordinary human nature has had some prominent defenders. For instance, philosopher-theologian Thomas Aquinas regarded humans as having some inchoate, implicit understanding that underwrote their thinking about God. He wrote, "To know in a general and confused way that God exists is implanted in us by nature,"[24] and, "There is a certain general and confused knowledge of God, which is in almost all men."[25] Akin to Paul's writings concerning the law being written on the heart (e.g., Romans 1) and Aquinas's notion of a divinely implanted knowledge of God, Jean

[20]Boyer, 2001; Barrett, 2011.

[21]Barrett, 2012; Kelemen & DiYanni, 2005; Järnefelt, Canfield, & Kelemen, 2015.

[22]For example, Norenzayan (2013).

[23]See, for instance, Clark and Barrett (2010), Greenway and Barrett (2018), and, for a different but resonate approach, Green (2015).

[24]Aquinas, *Summa Theologiae*, I, q. 2 a. 1, ad 1.

[25]Aquinas, Summa Contra Gentiles, III, 38.

(John) Calvin wrote that a "sense of Deity is inscribed on every heart."[26] Calvin went on to argue that religion cannot be an invention of humanity, for such an invention would have no place in the human mind were it not for some sense of the divine existing in the mind already. Even those who deny the existence of God find themselves wondering about His existence from time to time despite their resistance.[27] Indeed, Calvin wrote, "This is not a doctrine which is first learned at school," and such belief is "one which nature herself allows no individual to forget, though many, with all their might, strive to do so."[28] The cognitive science of religion appears to suggest something similar: the tendency for groups of humans to make sense of the world and their experiences in it in terms of some kind of God or gods is a natural extension of the way human minds work in community.

The sciences appear to tell us that belief in some kind of god is natural, but it does not follow from the sciences that we ought to form a loving relationship with a god or specify which god is the right one to love. We need something beyond science to tell us these things. Tyler Greenway and I have argued that Christian theology, especially concerning Jesus, addresses these issues.[29] It is God, showing who He is through Jesus, who helps us better identify the specific properties of the God who is worthy and willing to enter into a loving relationship with us. It is Jesus' life, death, and resurrection that can give us motivation to receive God's offer of such a relationship.

Exercising care for God's creation. What about the theological proposal that humans should function as God's stewards or representatives in living out our vocation to care for the rest of creation and thereby exercise dominion over the plants and animals? Here, too, the evolutionary analysis provides useful insights. Of course, the human ability to acquire and use tremendous amounts of information that varies from

---

[26]Calvin, Jean. *Institutes*, I.3.1.
[27]Calvin, Jean. *Institutes*, I.3.2.
[28]Calvin, Jean. *Institutes*, I.3.3.
[29]Greenway & Barrett, 2018; Greenway, Barrett, & Furrow, 2016.

place to place, problem to problem, and individual to individual (as reviewed in chapter five) is just the bag of tricks we need to wisely care for each other and the rest of creation. Each environment has a different geography and ecology of plants, animals, and other organisms that interact in unique ways. It takes building up local knowledge to understand how humans can find a place in such a niche without reconstructing it in a deleterious way, but maybe even facilitating the thriving of the other organisms in it. Do we humans (mis)manage the world? Clearly, we do to a certain extent. Throughout this book I have noted that the three big clusters of distinctively human capacities are the ones that enable us humans to solve fitness problems in distinctively human ways, but they are also the means by which humans change the environment around them in dramatic ways. We will impact the rest of the created world whether we want to or not, but we should be more careful to have more modest impact. Other organisms are not unlike those kids trying to reach the platform that previous swimmers have pushed away if they are not careful.

That said, ancestral human interactions with plants and animals as food sources and as sources of clothing, shelter, labor, and companionship have importantly shaped the kind of animal we have become, the kinds of societies we have developed,[30] and the purposes we have pursued. As described previously, the development of our big brains may have been possible only as a result of consuming animals as a significant part of our diets. Living in cold northern latitudes or high up in mountains would not have been possible without certain plants or animals falling under partial human control. Even if often abused, exercising "dominion" through selective breeding of plants and animals may have been critical to us becoming human in some important senses. The original and subsequent audiences of Genesis and the Psalms likely lived in an agrarian context, surrounded by domesticated plants and animals—species that were somewhat under the control of humans. Possibly, the idea that to

[30]Diamond, 2003.

exercise "dominion" in God's place and to be human are closely intertwined would have struck these audiences as reasonable.[31]

Here again, evolutionary psychology gives us additional reasons for thinking theologians have been on the right track in the sorts of properties and functions they have regarded as critical for understanding human nature, but evolutionary psychology does not and cannot tell us what to do about it. Christian theology, however, can cautiously affirm the role for humans as stewards or caretakers of the natural world: we can and will "rule" over the plants and animals, but whether we do so under God's authority and as His representatives with His aims in mind or by our own aims and thereby usurp God's authority is a different matter. Fortunately, if the *sensus divinitatis* is part of human nature, we will tend to have thoughts about one or more beings outside nature that may have been involved in creating or ordering the world. It is a modest inference to suppose it may be best to interact with that world in a way that meets the approval of the divine.[32]

TELOS REVISITED

Evolutionary psychology can give us insights into the specific features of human nature and how they may solve particular problems that face us, but it will not tell us how we should use that nature to address the nature-niche gap problem that hinders thriving. Evolutionary psychology also does not and cannot tell us what our purpose is and what we should do. If thriving is moving toward one's telos or purpose, evolutionary psychology can give us useful information and fresh perspectives on what it means to be human but cannot tell us what our general human telos is, let alone any personal telos. To answer these questions, we must turn to a prescriptive, normative system such as a philosophy or theology, ideally one developed with use of the appropriate sciences.

---

[31]Barrett & Greenway, 2017.

[32]Indeed, even people who do not recognize a divine cosmic creator often believe in lesser gods or spirits who play a role in ordering nature. Abuse of this natural order or failure to perform the right rituals in interaction with these spirits may lead to calamity (e.g., see Barth, 1975).

Some interesting resonance can be found between Christian theological anthropology and the evolutionary study of human nature.[33] Both highlight some special human features that seem to set us apart from other animals, including our ability to reason and exercise rationality, use of language, freedom and morality, and relationality. As noted above, each of these has been identified with the *imago Dei* (image of God), but the *imago Dei* has also been identified with two possible functions or roles of humans: to love God and others and to care for the natural world as God's representatives. Our analysis from evolutionary psychology also affirms these two functions as natural extensions of human properties. It even appears that theology and science can agree that humans naturally have something like a *sensus divinitatis* (sense of the divine) that may help prompt us to both enter a relationship with God and to interact with the natural world bearing in mind that the natural world was ordered and created by someone else.

Human telos, and hence human thriving, are importantly anchored in loving God and loving others and representing God in caring for His creation. Our submission to God's authority and demonstration of love to Him and those beings—human or not—that God cares for are how we become more Christlike. Such general claims, however, were already available from Christian theology, clearly highlighted by God's command to Adam and Eve to care for creation (the cultural mandate, Genesis 1:28) and the greatest commandment to love God and each other (Matthew 22:36-40). The resonance of evolutionary psychology with these theological claims may give us reason to think that evolutionary psychology is not antithetical to theological anthropology, but has it given us anything new and useful? I think so, and I put this conviction to work in the final chapter.

---

[33] I am not, however, arguing that there are not some points of friction or that there is a glove fit between what each and every evolutionary scholar says and what each and every theological anthropologist says. There is too great a diversity of opinions in both camps for such strong convergence to be possible.

8

WHAT IS YOUR TELOS?

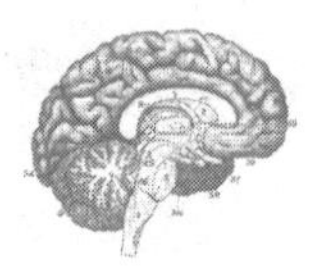

For theological reasons, one may think that caring for an infirm friend is the right thing to do, but in striving to actually meet the friend's needs and restore them to health, one will likely be much more effective with insights from modern medicine and nutrition. These science-based fields help a caregiver know what will assist in their goals and what actions or situations may work against their goals. Similarly, perhaps contemporary science can also give us tools for more effectively channeling our efforts to thrive.

When integrated with Christian theology, evolutionary psychology gives us a new set of tools for looking at what humans are and are not good at and why, along with examining our natural tendencies and proclivities. These tools may not tell us directly what we should do in order to thrive, but they can give us insight concerning the consequences of proposed actions. Furthermore, they may help us anticipate and overcome challenges to thriving and proactively plan and pursue a thriving life. This science-theology integration is in its infancy and I do not pretend to have solved the problem of human thriving. Rather, my aim is to open a new chapter in the study and pursuit of thriving by giving some examples of how thriving may be more effectively illuminated by bringing evolutionary psychology and Christian theology together than either field can do on its own.

PURPOSE AT THE SPECIES LEVEL

As discussed in chapter seven, humanity as a species appears to have at least these dimensions of a God-given telos: to love God and each other and to represent God's wishes and aims for the natural world through our care for it. If God has created us with the unique bundle of capacities that make us human so that we can love God and each other eternally and so that we can represent God's reign over creation on earth through our vocations, then surely that is reason enough to engage in these purposes. Human thriving, generally and typically, means faithfully pursuing these functions and becoming like Christ.[1] Nonetheless, our own experiences and evolutionary psychology remind us that we are not gods but creatures. We are "dust of the earth" and animals descended from other animals. To pursue our theologically described telos, we must do it with the toolkits we have available to us in the environmental niches that face us. Our God-given telos will be shaped and informed by our God-given nature in interaction with our God- and human-shaped niches.[2] Treasure in earthen vessels, indeed (2 Corinthians 4:7).

Loving others. Consider the act of loving others. This purpose is at once simple and intractable. Even people we find easy to love most of the time are sometimes hard to love. What, then, about those people we don't even know or who are our enemies? I don't think I am alone in feeling like loving others is no trivial task, and so it seems that God has put a steep climb in front of us if we are supposed to love others as we love ourselves. Yikes.

Why is loving so hard? The answer depends on what is meant by *love*. Many Christian thinkers have noted that our English word *love* is used to signify many different things. For instance, in C. S. Lewis's *Four Loves* he describes four different kinds of love: *storge* (affection, as in a parent's

[1]This synopsis concerns humans generally, as a species. But we are individuals within this species, and, as such, we carry limitations and gifts that are unique to us due to our personal "nature," development, history, and situation. These impact the particular roles we can play in God's redemptive and restorative work in this world, our vocations. But this diversity of personal capacities should not lead us to lose sight of the common, unifying features of being a thriving human.

[2]King & Whitney, 2015.

affection for a child), *philios* (friendship), *eros* (romantic love), and *charity* (selfless or unconditional love).[3] Lewis regards the four loves as varying in how uniquely human (among creatures) they are and how difficult they are to express well. To develop just one of the four, storge is the label (from Greek) that Lewis uses for the kind of affection that seems to characterize the relationships in families and sometimes between acquaintances. It is a love that isn't earned or conditioned on the qualities of the beloved. Parents love their children because they are theirs and not because they have qualities that make them loveable. This is storge. Lewis regards storge/affection as the most "natural" love and sees versions of it in other species; it is a good love that we should exercise. And yet in Lewis's analysis, affection can go wrong. It can lead to smothering the beloved or enabling poor decisions and conditions in the beloved. Storge can also have a self-serving quality about it, as when parents want their children to remain dependent on them so they can continue to meet the children's needs. This is storge gone wrong. Lewis draws out similar weaknesses in philios/friendship and eros/romantic love, but he also sees precursors of these types of love in other species and a very natural place for each in humans. These are loves that are naturally part of the human experience even if at times they are challenging to exercise well. In contrast, Lewis regards the greatest love, charity, as a supernatural gift, one humans are unable to participate in but for the grace of God. It does not come "naturally."

Philosopher Eleonore Stump, developing the thought of Thomas Aquinas, provides a similar but more detailed treatment of this highest love, and her analysis in interaction with psychological science makes clearer why charity is unseen outside of humans and hard for humans to express.[4] Stump observes that for Aquinas, love has two components. It requires "desire for the beloved those things which in fact contribute to the beloved's flourishing."[5] Second, it requires a desire for union with the

[3]Lewis, 1960.
[4]Stump, 2006.
[5]Stump, 2006, p. 28.

beloved, but this union has to be "a kind which is compatible with what is in fact conducive to the flourishing of the beloved."[6] It is no good to want the beloved to have what the beloved wants if there is reason to believe that the beloved's desire is misplaced. It is not enough for the lover to want for the beloved what would be good for the lover because those goods may not be what the beloved needs to thrive. And if being united with the beloved would be bad for the beloved's flourishing, that desire should not be acted on. An example of this kind of high love might be the mother who tearfully kicks her adult child out of her home because she knows he has become unhealthily dependent on her and it is best for him to not be with her right now.

Such a love as this will not be realized through brute instinct or a simple behavioral routine but requires strong contributions from all three of the distinctively human bundles of capacities. This love's dependence on our very special social nature is clear. It is about one person desiring and acting to bring about the flourishing of another specific individual, which requires distinguishing that person from the herd and discerning his or her needs. Perspective-taking, empathy, and higher-order mindreading will be required to do it well. What are this individual's experiences, feelings, and thoughts? What are the beloved's current desires and aims and how can I facilitate their satisfaction? This love also demands (unless it is harmful to the beloved) the formation of an interpersonal relationship or attachment.

Loving well also requires culturally conditioned expertise. The lover must distinguish what would enhance flourishing of the beloved in his or her situation, place in development, and niche. What would contribute to the beloved's thriving or flourishing may be very particular to that individual and discovered only through use of information acquired from others and built on cumulative culture. This demand on culturally acquired expertise may be especially great when we try to love those who do not share our local history or the values that we automatically take for granted.

[6]Stump, 2006, p. 28.

Importantly, this high sort of love requires a healthy dose of self-control, too. To love well, it seems, we have to check our impulses to give others what they want just because they want it. We must override our temptation to give them what we would want, too: just because I might want something if I were in that situation, it does not mean that is what my beloved wants or would be best. Of course, our self-control must also be exercised to keep us from distorting our care for the other into an indirect way to satisfy our own needs and desires. Indeed, love of this sort requires me to be willing even to sacrifice relational union with someone I cherish if that would be best for the beloved. To love well requires unselfishness and so we must use that peculiarly strong human ability to deny ourselves.

To love in this way well, the wholeness of our distinctively human nature must mobilize properly.

To love in this way well, the wholeness of our distinctively human nature must mobilize properly. To love in this way consistently, we will have to use the capacities that enable our sociality, our use of specific information, and our self-control, and we will need to practice loving in this way until it becomes easier and more fluent.[7]

Our human nature makes it possible for us to express this high form of love, but often it carries liabilities that make it challenging. For example, our attachment system helps us form intimate bonds with caregivers in infancy. This attachment relationship is complex and important. Babies are not only naturally prone to be oriented toward and receptive to their caregivers, but the quality of that relationship is also a function of cues received from the caregivers. Additionally, these early relationships can lead to security or insecurity in our future relationships. They inform our relational styles. If, as many theorists have suggested, these styles are anchored by early interactions, then parenting characteristics may leave marks on how we love others for the

[7]My discussion here is indebted to Michael Murray's essay (2019) and subsequent exchanges with him.

rest of our lives. For this reason, how successful we are at loving others may be importantly influenced by how we were first loved.[8] These early contextual factors that lead to difficulties in loving others well may be labeled as sin, but so labeling them does little to help us understand or solve the problems they create. The psychological science, however, helps us see that effective early parenting and therapeutic interventions to help people improve their attachment styles later in life are both methods of helping others love better and thrive.

If we humans naturally have social networks with rings of decreasing intimacy (see chapter four), then how we love will be shaped by those rings. It may simply be beyond human capacity to love 150 individuals as deeply and responsively as we do our more intimate circle. Likewise, it may be impossible for most of us to actually have intimate, personal, loving relationships with more than 150 people. We cannot literally love each and every person in any prolonged way.

What about the notion that we should love all people equally? Evolutionary psychology suggests that to do so would be impossible and, thus, should prompt us to revisit Scripture to see if God has really called us to love each and every individual to the same depth. In this case, the science suggests we can't. Here evolutionary psychology gives insight into what this biblical mandate might mean and how it might be lived out. We could exercise a high degree of beneficence toward strangers and mere acquaintances, but perhaps that is the extent of our capacity to enduringly love those outside our social network of one or two hundred people. This is important. God sometimes asks of us things that require us to rely on His strength and provision, that we cannot do alone. But it would be unusual for God to commonly command us all to do things that fall outside of our fundamental human nature. What damage could that do to our relationship with God and our motivation to do what God asks of us? Evolutionary psychology provides insight into human relational capacity, which in turn helps us understand the biblical mandate to love

[8]An interesting theological parallel is suggested in 1 John 4.

one another. The scientific perspectives should, at the very least, make us cautious about some interpretations of what God asks of us. In this specific case, though God has commanded us to love all people, it seems unlikely that He has commanded us to love all people equally.

Even if we are not reaching to the heights of charity, love minimally requires that we regard even strangers and enemies as having moral worth. In chapter six I described moral foundations theory and the sorts of moral intuitions these foundations appear to generate: intuitions concerning care/harm, fairness/cheating, loyalty/betrayal, authority/subversion, and sanctity/degradation. I noted, however, that the reach of those considered to have moral value, one's moral circle, might be fairly small by default. Our moral foundations may tell us it is wrong to harm members of our in-group, but our in-group may extend only to our clan or village. For example, in his famous ethnography of the Baktaman of Papua New Guinea in the early 1970s, anthropologist Frederik Barth notes with little surprise that the Baktaman ate a tribe member's mother-in-law because she was from a different village, something they would never dream of doing to one of their own people.[9] Likewise, the news, world history, and ethnography are full of accounts of people who tenderly love and care for their own while committing frightening acts aimed at out-group members and being congratulated for it when they get home. Our moral circles can be very small, indeed. A small circle may be the natural default condition.

Unlike social networks, however, I know of no research that suggests that people cannot expand their moral circle to include more and more people—and even animals—as having moral worth. Our nature supplies only a starting point, but cultural factors, including religious traditions and the disciplines they encourage, may help grow us in our ability to love others. Our theology affirms that recognizing the moral worth and God-given dignity of all persons is part of our telos, and in this case evolutionary psychology helps us see the obstacles we are up against to live

---

[9]Barth, 1975.

out this aspect of our telos without suggesting that surmounting the challenges is impossible.

These three examples (attachment, social networks, moral circle) illustrate three ways in which sciences—including evolutionary psychology—can help inform our theological understanding of what we are up against and what psychological resources we have when trying to love well. And if loving is part of our human purpose or telos, getting it right is central to thriving. Concerning attachment, the science can at once show us how environmental factors can push our thriving off course while also helping us find ways to prevent and remedy these factors. The social networks example shows how natural parameters on our thriving may have limits—they are just part of human nature—but we may be tempted for theological reasons to try pushing past them. The case of our initially limited but perhaps expandable moral circle illustrates the possibility that human nature may tend toward very imperfect thriving but that it can be improved upon. In all three cases, it is an integration of theology with the sciences that bears the most fruit and not either one on its own.

Loving God. When it comes to loving God, our theology may tell us that only an unchanging, wholly good, and eternal God is worthy of our unswerving and unconditioned devotion. Nevertheless, as with loving others, our human nature goes and makes things complicated. Again, theology calls this "sin" and sometimes our "sin nature" and in doing so expresses some important truths about our situation. As a tool in our theological toolbox, evolutionary psychology can extend these insights. Because of humanity's general separation from God, we often form our attitudes and conceptions of God's character through relationships with humans. Consequently, the quality of our relationship with God can be distorted through some of the same mechanisms that limit and distort our relationships with each other. We may find it hard to form a secure relationship with God because of our general attachment difficulties. Our relationship with our earthly parents, perhaps especially fathers, may flavor our relationship with our Heavenly Father, for good or ill.[10] Likewise, because we grow so accustomed to thinking about human capacities, mental properties, and fallibilities, we may unintentionally or even unknowingly extend these to God, what has been termed "anthropomorphism."[11]

As sketched in chapter seven, cognitive science of religion has begun identifying how natural cognitive systems may incline us toward the notion that at least one higher power, or god, exists, perhaps especially one who has been involved in bringing about the apparent design and purpose we see in the natural world.[12] Though our nature may incline us toward such beliefs, our attitude toward this god (or any others) is undetermined. Human selfishness may prod us toward treating gods instrumentally instead of appropriately showing deference, respect, awe, and even gratitude.[13]

People can and do also misrepresent God's character and intentions for us to each other, sometimes unknowingly but sometimes deliberately. Our social learning biases (described in chapter five) may make us especially receptive to misrepresentations that come from people of high prestige whom we perceive as relevantly similar to ourselves. If such ideas take hold in a group, the tendency to conform may help them become the norm and subsequently difficult to resist. Ordinary natural psychological dynamics that usually help us solve life's problems can be hijacked to distort the most important component of human thriving.

The news isn't all bad. It appears, too, that basic ideas about the existence of at least one superhuman being who helps account for why we perceive design and purpose in the natural world and who is readily understood to be superknowing and immortal is an easy product of our natural conceptual endowment.[14] As mentioned in chapter seven, a body of scientific research supports something like a *sensus divinitatis* or sense

---

[10]Granqvist, Mikulincer, Geertz, & Shaver, 2012; Kirkpatrick, 2005; McMurdie, Dollahite, & Hardy 2013; Vitz, 2013.

[11]For an experimental study of anthropomorphism in God concepts, see Barrett and Keil (1996).

[12]Barrett, 2012; McCauley, 2011.

[13]See Green (2013). Adam Green nicely develops how a cognitive science of religion approach to Romans 1 offers a fresh theological perspective on natural knowledge of God.

[14]Barrett, 2012.

of the divine. So, instead of starting from scratch in building and understanding a relationship with God, humanity may have the basic raw materials from which to build.

Likewise, once we understand the psychology of human learning, including those social learning biases, we are in a better position to harness those biases in a positive, productive direction to point people toward God instead of away from God. Various sciences, including evolutionary psychology, may then be resources for people and communities to better know and love God.

Caring for creation. In the divine mandate to love each other, we already receive a directive to care for part of the created world, the human part. However, our species-level purpose extends beyond humans. Biblically, this further responsibility first receives articulation in Genesis 1 with reference to human dominion over the other animals and plants. As discussed in chapter six, this "dominion" does not appear to be justification for humans to do with as they please but is part of what it means to reflect or image God on earth. That is, humanity serves as God's representatives in part by overseeing and caring for God's creation (Genesis 1:26). We must never forget that the world is God's and in God's created order we serve God—not ourselves—in exercising dominion.

The second biblical hint at this responsibility to care for creation comes from the second creation account in Genesis chapter 2. In Genesis 2:7, God creates the man from the dust of the ground, and in Genesis 2:8 God creates a garden. Then, in Genesis 2:15, we read, "The LORD God took the man and put him in the garden of Eden to till it and keep it." Why did God place the man (the adam) in the garden? For a purpose: to till it and keep it. The original audience of Genesis consisted of agrarians whose livelihoods depended on the right care of plants and animals, and they would have understood this purpose, this telos, very well. God was not just inviting but commanding this representative of all humanity (the adam) to join him in caring for what God made good.

Today it is easy for most of us to think of ourselves as completely separated from the nonhuman aspects of creation. More than half of the world's population now lives in urban environments that are more marked by keeping nature at arm's length or even crushing it than in tilling it and keeping it.[15] I once knew a young woman who, upon going away to college, expressed her great fear and discomfort at going outside because "there is just too much nature out there!" And this college was not in some isolated pristine forest but a thoroughly civilized college town. This reaction represents many urban dwellers' feelings about the natural world: it fills us with discomfort, anxiety, or even fear. The natural world with its plants and animals, rocks, minerals, landscapes, and weather patterns seems foreign to us. No wonder the directive to care for creation—arguably God's first commandment to humanity—is often ignored entirely. It seems irrelevant to us.

A second reaction to this aspect of our human telos can cause us to err on the other side. Some environmentalism, Christian or otherwise, can be so concerned about human misuse and abuse of the natural world that it regards any and all human presence in the natural order as an intrusion and imposition. Humanity's special role as God's chosen stewards with instruction to "till and keep" is ignored. Hence, humans may be regarded as evil intruders on an otherwise perfect world,[16] or human cultivation of plants and animals is regarded as unethical in principle. Biblical theology demands that humans not abuse God's creation, but it also demands that we care for it actively with recognition that it is God's and not ours, not passively "care" by leaving it entirely alone. Striking the right balance requires wisdom and understanding, some of which can come from the relevant sciences. The fact that our psychology runs the risk of getting out of step with our environments suggests the importance of going slow when it comes to modifying the natural world away from ancestral conditions. Let's keep our nature-niche gap as small as possible.

If we require more support for adopting a measured, caring interaction with the natural world, an evolutionary perspective provides

---

[15]United Nations, 2014.

[16]God calls His creation "good" in Genesis, but never "perfect."

such help. As I have observed throughout this book, humans, like other animals, impact their environments when they attempt to solve regular species-typical survival problems. We cannot do otherwise and so it cannot be the case that we ought to have no impact on our environment. We have also observed that humans exacerbate their own fitness challenges when they too radically change their environments. In other words, failure to pursue our God-given telos to serve as good stewards of God's creation—and thus, failure to thrive in this dimension—also creates obstacles to our thriving in terms of having a good nature-niche fit. God knew what He was doing in directing humans to be the guardians and cultivators of nature: it is natural for us to impact nature and good for us to do it responsibly. The unique human ability to gather niche-specific information and exercise self-control in putting that information to good use equips us to serve this role, but we do need to actively seek the experience and expertise that will train up our hearts and minds to be effective stewards of creation. We need to avoid being estranged and distant from the natural world of which we are a part. We should never find ourselves saying, "I don't like it. There is too much nature out there."

When humans "till and keep," we leave our marks on the world. Gradually, as with domesticated plants and animals such as bananas and corn, dogs and horses, the lines between the natural world and the human world become blurry. There is nothing wrong with that: humans are part of the natural world, too. Indeed, leading theories of human evolution stress that interaction with plants and animals—including eating them—and domestication of plants and animals have been a part of making us the humans that we are. Of course we will leave our marks. Key for this telos, however, is to recognize that the consequences of our care and cultivation are also God's. As when a factory manager makes changes to the product, the factory and its

We need to avoid being estranged and distant from the natural world of which we are a part.

products are still the factory owner's and not the manager's.[17] Likewise, some part of the world does not cease to be God's creation just because humans have had a hand in its form. What follows from this observation is important: the directive to "till and keep" or exercise "dominion" does not stop with our city limits; it is boundaryless.

Furthermore, as we observed earlier, the human niche includes our social and cultural niche, including its artifacts and institutions: arts, businesses, entertainment, farms, governments, schools, and sciences. These, too, are not ours but God's. We are to care for them as part of our vocation, to "till and keep" them, with a recognition that they are God's.[18] Failure to do so is failure to thrive.

FAMILY PURPOSE: SPECIAL AND GENERAL

We started this chapter with comments about how Christian theology and evolutionary psychology may be brought together concerning human purposes at the species level: what humanity as a whole should be about. When thinking about our purpose in life or telos, it is easy to think individualistically: what is *my* purpose? I close this chapter with some comments on individual telos and thriving, but before turning there, notice that both Christian theology and evolutionary psychology affirm that many human groups also have purposes, from families to nations.

Consider families. In some ways, the biblical narrative is a series of interlinking accounts of particular families forming a particular nation of people from whom the Savior of the world, the Messiah or Christ, would come. Noah's family was chosen as the family through whom God would rebuild his relationship with humanity. Abraham, Sarah, and their family served to establish a lineage of people who did not see God as one of many competing tribal gods but the one God to whom people should be

[17]Extending the analogy, my colleague Sarey Martin Concepción observes that if the manager changes the product very much, the quality of the work and even the properties of the factory may change.

[18]Burdett, 2020.

devoted. Moses and his siblings had the purpose of leading Abraham's descendants out of slavery in Egypt and to the Promised Land where a new nation would be born. David and his family would serve the function of making the Israelites a unified and formidable kingdom centered on the worship of God, as well as being the lineage from which the Messiah would be born.

Families have specific purposes throughout Scripture, but they also typically (but not unswervingly) serve a common purpose: to "be fruitful and multiply," to serve as the structure and context for children to be born and raised up to be in a right relationship with God. Evolutionary psychology is consistent with the basic insight that families, as a social mechanism, have arisen as a means for raising and socializing children into adulthood and even into their own child rearing. Though not unique in forming stable mating pairs, we humans do seem to be unusually consistent in using pair bonding and surrounding extended family as our primary mode for bearing and rearing children.[19] The multigenerational investment in each other that we see in human families has no real match in other animals. The absence of invested parents and multiple invested adults in the life of a child is a risk factor for poverty and antisocial behavior later in life.[20] Of course, the biblical view is not just that families are for making and raising babies—families are also for helping us understand what our relationship with God should be like. The Bible uses familial metaphors for teaching us how to think about our relationship with God. God is our Heavenly Father, fellow Christians are our brothers and sisters, and Jesus is the bridegroom of his bride, the church.[21] One purpose of families, then, is to reveal truths about how God loves us. A thriving family will embody these truths. A thriving family will also contribute

[19]Note that our focus here is on families and not couples. Pair bonding is found in other species and it may be that some species are more faithful than human pairs in mating for life. I am not suggesting that the telos for every couple is having children.

[20]Benson, 2006; Damon, 1997; Lerner, Lerner, Bowers, & Geldhof, 2015.

[21]It may be that this familial language helps us expand our moral circle. If we talk about unknown fellow Christians as "brothers" and "sisters," maybe we will think about them a bit more as such and extend familial affection to them (Atran, 2002).

to the thriving of its members, that each may grow in differentiated pursuit of their purposes while maintaining unity and a coherent sense of family purpose.

Note, too, that even though we have reason to think that some specific families have a special purpose that is peculiar to them—not all families are called to lead a nation out of slavery—these special purposes rarely replace the general purpose of being the place for children to be born and raised in a way that places them on a path to love God and others. When biblical families fulfill their special purpose, they typically show signs of moving toward their general purposes as well. It is rare (but not impossible) for a specific purpose to be fundamentally at odds with a general purpose.

Families are thriving when they are fulfilling their purposes. Among those purposes is to provide the members of the family a developmental environment that encourages each of their individual thriving. This relationship between family thriving and individual thriving, however, runs both ways. Part of being a thriving individual is to love one's family well, and that means being invested in the thriving of family members. The thriving of selves and families is reciprocal. The more the family centers on what it should be, the more individual members of the family can approach what they should be; the more the individual members approach the target of what they should be, the easier it is for the family as a whole to be appropriately centered. As Pam is known to say—telos "turns 'me' upside down. Me thrives when we thrive."

COMMUNITY AND CHURCH PURPOSES

Individuals do not find themselves only in families but in many other rings of social arrangements as well, such as neighborhoods, towns, and nations. For Christ-followers, the church, local and global, are key communities that we both contribute to and benefit from. As with families, these various social rings may be characterized as thriving to the extent that they are oriented toward and pursuing right purposes. One purpose for them is to facilitate the thriving of the communities that make them up. The global church thrives if it is advancing the kingdom of God, but a big part of successfully doing so is to facilitate local churches in successfully playing their parts in the kingdom of God, which in turn is partly a function of whether the families and individuals in those families are doing their best to image God and live in Christlike manners. Of course, a local church's thriving is reciprocally built on the thriving of individuals and families, and the thriving of local churches contributes to the thriving of the global church.

Jesus' parable of the talents offers perspective on this multilayered thriving. In Matthew's telling, Jesus explains that the kingdom of God is like "a man going on a journey [who] called his servants and entrusted to them his property" (Matthew 25:14). Three servants are entrusted with wealth—talents—from their master and expected to put that wealth to work in service of the master and his estate. The master then returns and demands an account. Two of the servants have doubled their master's money:

> Now after a long time the master of those servants came and settled accounts with them. And he who had received the five talents came forward, bringing five talents more, saying, "Master, you delivered to me five talents; here I have made five talents more." His master said to him, "Well done, good and faithful servant; you have been faithful over a little, I will set you over much; enter into the joy of your master." (Matthew 25:19-21)

One of the three servants, however, has buried his one talent in the ground and is condemned for doing so by the master.

This famous parable is about the kingdom of God or what the world is like under God's reign. And what is it like? It is a place where God has given His servants (His image bearers) resources (talents) to invest for the good of the estate (the kingdom), and good investment of them leads to both celebration and a greater share in collaborating with the master. The estate as a whole, as well as the individual servants, benefit or thrive when the individuals take what the master has given them and make good use of them in service of the master, not themselves. In other words, the kingdom of God is a place in which the whole world (God's estate or kingdom) thrives when individuals live into their divinely given purposes and not alternative purposes they may desire for themselves. The church is God's primary earthly mechanism for advancing His kingdom on earth, for developing, aligning, and supporting individual purposes such that both individuals and their social rings, extending out to the whole of the world, thrive.

Bringing in an evolutionary perspective, these social rings of families, institutions, churches, towns, and nations are the social and cultural niche in which individuals develop. If these social conditions change rapidly, particularly away from the crossculturally common conditions that human psychology is naturally tuned to, the nature-niche gap may grow and threaten individual thriving. Those in positions of influence in leading and shaping these social groups, then, may contribute to either facilitating or frustrating the thriving of people in their sphere. How can social structures respect and work with, instead of against, human nature?

In Homer's *Odyssey*, Odysseus proves his identity by shooting an arrow through twelve axe handles, a feat that usurpers to his estate cannot achieve.[22] Consider a similar metaphor here: shooting an arrow through various rings to hit a target. It will be easier to hit the bull's-eye if all the hoops align on the bull's-eye. The hoops represent the various rings of communities we are in. It will be easiest to hit the target when all the hoops align on each other and on the bull's-eye if they are close to the target. Furthermore, if the hoops are far from the target, tiny misalignments will be magnified and it may be impossible to shoot an arrow through the various hoops and hit the target. Likewise, the more closely our various communities of relationships resemble what a Jesus-oriented community is supposed to be, the easier it will be for us to each hit the target as individuals and thrive. Our potential for thriving increases, then, when the communities we are part of are also moving closer to Jesus and

[22]We thank Adam Green for reminding us of this literary episode.

centering on his example. Part of doing so is to acknowledge the natural social, informational, and self-control needs and capacities and limitations people have and to gently help them bridge the nature-niche gap. Individual thriving is facilitated by thriving families and local communities all aligned on God's purposes for us. The more these communities act as church—the body of Christ on earth becoming what it should be—the more they provide a developmental context for the thriving of individual members. Reciprocally, the more individuals take the gifts God has given them and invest them for His purposes in relation to their niches, the more those local communities live into being what the church should be.

Our potential for thriving increases when the communities we are part of are also moving closer to Jesus and centering on his example.

INDIVIDUAL PURPOSES AND VOCATIONS

It is clear that finding one's individual purpose in life is a major concern for many of us today, even if there is something historically and culturally peculiar about this concern (as discussed in chapter three). Though the focus of this book is not to provide guidance for how to discern one's purpose in life, I do wish to end with a few observations concerning how a theology-infused evolutionary-psychological approach may help us in the pursuit of our personal, individual thriving.

Biblically and theologically, we must allow for the fact that sometimes individuals have a purpose that does not enhance their inclusive fitness. The easiest example is Jesus: he sacrificed himself for all of humanity without ever fathering or rearing biological children and he made disciples of his brothers who, in some cases, became martyrs for their commitment to him. Jesus, then, appears to have chosen a purpose that was not fitness-enhancing. His example clearly demonstrates that a simple equation between fitness and each and every individual telos cannot be made.

With that acknowledgment in mind, it would be suspicious if a disproportionate number of individual purposes were hard to square with our species-general purpose, with individual fitness, or both. It would be odd if God selected humans to image Him because of their nature that He created but then commonly demanded they do things that run contrary to that nature's general characteristics. In the beginning of Genesis, we read that after God's creation of various types of animals, He calls them good. When He creates humans, He calls us very good. He does not say, "Well, I guess they'll be very good if they can get over their nature." We have the features we have in great part because they have contributed to our fitness. These general characteristically human features, then, serve as heuristic guides to the kinds of lives we should pursue.

For instance, it would be surprising for God to demand that many people live as isolated hermits or learn to survive on their own without the benefit of others' expertise, as these conditions would run against our sociality and our great reliance on culturally transmitted information. We should also be suspicious of proposals that one's thriving is dependent on conditions that are so culturally scaffolded as to not be available to most humans throughout the ages, conditions that demand us to have resources far outstripping our nature. At best these special cultural conditions are means to ends but not ends in themselves.

To illustrate this tension, consider athletics. As noted earlier, the number one "spark" or purpose that teenaged American males cite in their lives is a sport or athletic pursuit. Imagine, then, if a third of adolescent and young adult males really did make sports the primary purpose of their lives. It is not hard to see that such a scenario would be economically unsustainable and lead to a huge number of physically spent thirty-five-year-olds with bad knees and little preparation for living out the remaining fifty years of their lives. Given that professional sports is a culturally peculiar way to best manifest humanness, it would be surprising if God actually made many (if any) individuals' central purpose to be a professional athlete. Before you think I am picking on athletes, the same could be said for professional scientists or scholars. I can be a scholar only because of very peculiar cultural conditions and a degree of societal wealth that is not available to everyone. Am I willing to take my own medicine and say that my ultimate purpose is not to be a scholar? Yes.

My analysis suggests that the purpose I need to pursue to have a thriving life is not fundamentally about being a professional scientist or scholar (or athlete, or pianist, or bus driver, etc.). I will thrive when I am part of a thriving community in which I try to live a life imitating Jesus' focus on loving and serving God and loving others. That is my primary purpose. I may have a secondary, vocational purpose that is culturally conditioned (in my case, to be a scholar) and many possible means by which I live out my primary purpose. My secondary purpose may change over time. Mistaking the means for thriving (being a scholar in the service of God) for my primary aim could be disastrous. How? It could lead me to neglect loving others well—placing my scholarly achievement above other people—which would erode the love and support from others that I need. It could entice me to prioritize my vocation over appropriately contributing to my family and my community. As I form my collegial relationships, I could reduce these individuals—each created to image God—as mere sources for information or aids for spreading "my" ideas. The result of these cumulative misdirections could be loneliness, anxiety, and a life poorly lived.

Consider also a boy who seems perfectly suited for playing American football.[23] He is considerably stronger and larger than other boys and most grown men, so big that his youth tackle football league forbids him from running the football because of concerns for safety and fairness to the other kids. He seems a prime candidate for someone "made by God" to be a football player. And then, in high school, he suffers a career-ending injury. It is all over. Does that mean he can't have a thriving life? Of course not. But if his community has consistently told him that he is "born to play football" and he has adopted that as his ultimate purpose in life, he will likely be confused, deflated, and perhaps disappointed in

[23]This case is based on a true story.

life and angry at God. There is danger in anchoring one's thriving to something that is so particular and contextual. An injury or change of situation can take it all away and be followed by depression, anxiety, and feeling lost. But I believe God doesn't make our purpose or thriving dependent on anything so environmentally contingent or so easily taken away. Good things become suspect when out of proportion with basic human nature and needs and basic human God-given purpose. As Pam says, thriving is more contingent on who we are becoming than what we are doing.[24]

Thriving is more contingent on who we are becoming than what we are doing.

For these reasons, it is critical to distinguish one's human purpose or telos from one's personal current purpose or vocation. Vocation is a helpful construct for focusing one's understanding of a specific, personal way to steward God's creation, the human and nonhuman parts of it; our telos is largely shared with others, enduring, and context-free. Our vocation is personal and situational. It may change as we change and our situation changes. It affords the opportunity to reflect on how we are each particularly called to tend God's creation while becoming more Christlike. That is, an appropriate vocation must align with one's telos.

Makoto Fujimura suggests that we engage in "culture care" in parallel to "creation care."[25] He emphasizes the potential for promoting flourishing and thriving through constructive engagement with art, business, farming, and all aspects of culture. This paradigm is rooted in an appreciation of all forms of culture and industry as a part of God's sacred creation. From this standpoint, vocation provides the opportunity to participate in God's ongoing work of creation, redemption, and perfection of this world. Though we may each have an individual vocation, that vocation is informed by a telos that is not fundamentally about

---

[24]King, in press.
[25]Fujimura, 2015.

ourselves as individuals or what is best for us, as if we could be isolated from other humans, the rest of creation, or God.

When attempting to discern our own purpose and pursue it, thereby getting on a thriving path, Pam and I suggest the following questions as ways to test out a possible vocation or to see it more clearly. These questions are a means of sharpening your vision or understanding of your vocation. Pam offers a first set of questions to help you focus on a general sense of vocation that is suited for you based on the formulation of telos, summarized in chapter seven. Then I offer a second set of questions based on issues raised by evolutionary psychology in this book that I hope will help fine-tune your sense of vocation.

I do not regard yeses or noes to all of these questions as necessarily decisive but offer the questions merely as touchstones. When a possible vocational purpose comes to mind, ask yourself . . .

What are your greatest strengths, gifts, and competencies? What are you passionate about? Focusing on yourself as a particular person with a unique set of gifts and interests, which of your strengths offer you the deepest joy when you exercise them? What spiritual gifts do you find meaningful to practice? What competencies or other strengths are you aware of?

What are the greatest needs in the broader world around you that ignite you? When there is alignment between your gifts and passions and the local or global needs around you, consider that there may be a meaningful convergence of vocation and telos.

How does pursuing your gifts as they align with the world's needs enable you to become more like Christ? If Christ is the perfect image of God, then a shared telos of humanity is to become conformed to Christ. Conformity does not mean uniformity. We are to become like Christ as ourselves. At a very practical level, I ask how you are called to become like Christ at this season in your life. What virtues are you called to practice or nurture? How can you pattern your life after the life of Jesus? How does your emerging vocation help or hinder this journey of becoming more like Christ (i.e., your telos)?

The following questions are more specifically based on our discussion of evolutionary psychology. When considering a potential vocational purpose, ask yourself . . .

Does it help you love God better? Loving God is the central telos of any human life. If a proposed vocation is not an extension of or aid in this telos, then rethink committing your life to its pursuit.

Does it help you love others better? Similarly, if a proposed purpose interferes with us loving others—perhaps it requires us to put ourselves first over others instead of loving others as ourselves—then proceed with caution. We should be in the business of putting people before careers, economics, entertainment, politics, sports, and most everything else, including institutions. Those inner rings of our social networks—those fifteen best friends and closest family members—are sacred gifts for us to care for and be cared for by. Even within the church we can ask whether certain communities or groups effectively create space for us to be loved and to reciprocate that love. Whether a group is characterized by love is more important than whether it is fun, entertaining, educational, economically lucrative, or even "Bible-focused."

Will it help you grow your moral circle? Of a piece with the previous question, one motivation for pursuing a particular vocation is that it helps us love more people by expanding our moral circles. Some pursuits, such as joining a political party or movement or working for a franchise or corporation in a competitive field, may have virtuous aims associated with them, but you may find that the means to that end may lead you to carve up the world into sharp in-groups and out-groups. You may grow very fond of your in-group—your sports team, your political party, your military unit—but come to devalue or even dehumanize everyone else as a result. Is your primary in-group characterized by loving not only the in-group but also those outside it? Perhaps you already know you have difficulty loving certain out-group members and ascribing moral worth to them. Perhaps a vocation you are considering would create opportunities for you to learn to love people of a different culture or social status whom you find difficult to love. If so, that fact may count in its favor.

Will it make you part of a thriving group? We cannot fully thrive as individuals if our group is not thriving; our group cannot thrive if the individuals in it are not thriving. Freedom and encouragement to pursue individual purpose must be in fruitful and virtuous tension with group purpose. For instance, a family that invests exclusively in one member's dreams, whether a parent or a child, is not thriving. Sometimes it is good for a group or family to support one key activity that becomes identified with one member of the family more than the others, but group thriving cannot be sacrificed for any individuals or vice versa. "Me thrives when we thrive."

Will it involve treating cultural products or institutions as belonging to God but entrusted to you to care for? Sometimes our vocation has more to do with our attitude, values, and approach to a pursuit than the pursuit itself. One could run an excavation company or a restaurant as a divine calling if doing so is properly motivated and executed.

Will it reduce your estrangement from nature? To effectively serve as stewards of God's world, we need to know it better. Partly, that can be achieved through study, but a big part is to experience it, to relate to it more intimately. It is through our interaction with nature that we survive and thrive, and our interaction with our natural niche impacts our nature and our thriving. We are creatures and are constantly shaping it and are shaped by it. Are we doing so responsibly and with a recognition that, as the old hymn says, "this is my Father's world"? Or are we impacting the world around us to pursue transient personal goals, to enjoy our own gifts—creative abilities, physical abilities, and minds—as ends in themselves?

THRIVING WITH A STONE AGE MIND

Ending with questions for focusing one's personal thriving comes with a risk. If this book's analysis has been helpful, it has demonstrated that thriving is not merely an individual concern but presents tensions among individuals, communities, and cultures. Furthermore, our earthly existence confronts us with tensions among humans, other species, and our planet. While recognizing our own share in human nature—including its distinctive gifts, limitations, and tendencies—we each need to discern our own personal and God-given way to bridge the gap between our specific nature and our niche. Yet our community and our culture shape both our resources for bridging the gap and the characteristics of our niche. Furthermore, our own bridging efforts will change the gap for others. Will we help others close their gaps as we bridge ours? Will we help create a cultural and physical environment that has a smaller gap for future generations? Our personal thriving is not merely an individual concern; to truly thrive we must help others thrive.

The analysis presented here suggests four errors to avoid as we attempt to thrive, the first two focused on our nature and the second two focused on our niche. First, we could blindly ignore the fact that we as humans have a particular nature that grants us strengths and limitations that shape how we thrive. We are not gods but creatures with a toolkit for doing life in this world. The situation is like having one end of a bungee cord anchored to the ground and the other end tied to us. The nearer we are to the anchor spot, the more freely we can move; the farther away we are, the more tension we will feel and the more support or effort it will take to get where we want to go. Forgetting this fact may produce great frustration as our creatureliness interferes with our grandest visions for ourselves and for others. A second error would be to view our nature rigidly and deterministically. Sure, our Stone Age minds incline us to think and act in certain ways, but part of our human nature is to stretch and adapt to new situations through the support of others, through learning, and through cultural innovation. The bungee cord is real, but it is not a short, rigid tether to one spot. It is long and springy with lots of potential for creativity and movement. Just because we are inclined to act in a particular way does not mean that we have to do so or that it is good for us to do so.

We are not gods but creatures with a toolkit for doing life in this world.

Understanding our nature gives us a good grounding for discerning how to address our niche and the gap with our nature, but we need to understand our niche as well. Here, too, we risk two opposing errors. The first error would be to focus only on how to get ourselves across the gap without recognizing our ability to change the niche and, hence, the nature-niche gap for others. Remember that image of children trying to swim from a pier to a floating raft? If the raft is not anchored, and if the first child who tries to get on it thrashes around too much while swimming toward it, the approach can actually push the platform farther away. The result is that the next child has a greater gap to cross. Similarly, if we focus only on our own efforts to cross the nature-niche gap, we may actually make things harder for those who come behind us, other humans as well as other living things. Insofar as part of our telos is to care for each other, making it more difficult for those around or after us to thrive does not promote our thriving either. Being too focused on our own efforts to cross the gap may prove counterproductive. Rather than push the niche raft further away by our approach, we should be concerned with how we can pull it closer. There are two ways to thrive: move across the gap or close the gap. How can our efforts change our physical and cultural environment so that the gap is narrower for others? How can we be agents of transformation and reconciliation in this way?

Of course, the other side of this error is a risk as well. Think of the child who is so afraid to swim to the platform that he does not even try but demands the raft be brought to him. Such a child will not enjoy the swim, will not become a stronger swimmer in the effort, and will never experience the joyful satisfaction of getting across the space. Gaps give us the chance to jump and grow. It appears that part of being human—part of our thriving—is to work with others in venturing across the chasm between our nature and our niche. We are niche constructors. We can't help that, and we can't live in some kind of romantic primitivism as if we were orangutans or bears. We will use our social, informational, and self-controlling nature to bridge the gaps in our niches. It is this very quest throughout human history and prehistory that has made us what we are.

It may, in fact, be a big part of what it means for us to be made in God's image.

To sum up, evolutionary psychology reinforces a familiar theological theme: that we are creatures, not gods. Who we are and what it means for us to live an abundant life is importantly shaped by how our nature interacts with our niche. This niche is one that God has graced us to impact. But how will we impact it? Thriving requires that we marshal our nature to impact our niche as God has directed: to love God, love each other, and care for creation, including those aspects of creation that have been importantly shaped by human activity. Knowing our nature better by bringing together theological insights and those from relevant sciences, including evolutionary psychology, will give us new tools for effectively pursuing the telos to which God has called us. In this way we can begin the journey of thriving. By the grace of God, God will bring our thriving to completion.

AFTERWORD

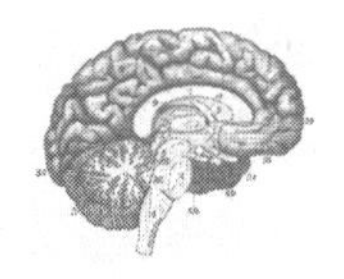

In 2012, the BioLogos Foundation launched its Evolution and Christian Faith program. This program funded a number of projects that explored how one might see harmony between science and biblical Christianity, and we (Pam and I) were fortunate enough to be one of the awarded teams. Together with our colleagues in Fuller Theological Seminary's Thrive Center—Jim Furrow, Sarah Schnitker, Oliver Crisp, William Whitney, and Tyler Greenway—we began regular meetings to discuss how evolutionary psychology might illuminate a Christian understanding of human thriving. Considerable research exists concerning the conditions under which young people may develop character strengths and virtues, basic competencies, relationships, and a sense of noble purpose that will support them in becoming thriving people—not merely less miserable people. Nevertheless, we recognized that much of this research is highly situational, often reflecting contemporary North American urban contexts and challenges. We wondered whether an evolutionary perspective, with its emphasis on our commonalities with humans everywhere past and present, might help us discern broader patterns that could illuminate what human thriving is even outside of early twenty-first-century Atlanta, Chicago, Los Angeles, or Vancouver. Could evolutionary psychology help us better understand thriving in a way that would help us promote thriving in young people regardless of where they live and help us anticipate thriving challenges and opportunities the future might hold?

Yes and no. Through readings and conversations with each other and with numerous colleagues, including Kendall Cotton Bronk, Kutter Callaway, Matthew Colwell, Tim Dally, Bill Dyrness, Bob Emmons, Adam Green, Joel Green, Megan Hutchinson, Mary Helen Immordino-Yang, Matthew Jensen, Veli-Matti Kärkkäinen, Jason McMartin, Michael Murray, Darcia Narvaez, Jeff Schloss, and Kirk Winslow, we quickly came to see that evolutionary psychology provides some valuable tools for thinking about what it means for humans to thrive, but it had to stand in a broader meaning system to fully deliver on its promise. We thank these colleagues who have educated and challenged us.

This book represents the culmination of our Evolution and Christian Faith research project, and we are truly grateful to the BioLogos Foundation and the John Templeton Foundation for making the project possible. TBF Foundation and Fuller Theological Seminary's generous sabbatical program also created space for the writing and revising of this book. Further room for exploring, sharing, and refining ideas shared here were enabled by the John Templeton Foundation–funded TheoPsych program through which we had the opportunity to discuss psychological science concerning human nature with many gifted theologians and philosophers. The BioLogos Foundation's Conversations on Human Identity and Personhood project, with its workshops of scholars from various human sciences, philosophy, and theology, has also been helpful in improving our thinking on this topic.

This volume would also be much the worse if not for the input of a courageous group of students who gave us feedback on the manuscript during what may have been the first course in evolutionary psychology offered at an American seminary: Andrea Canales, Tara Fairbanks, Gillian Grannum, Aylwin Leung, Hannah Myung, Tom Paulus, Alexandra Scott, and Chase Yeatman. Rebecca Sok, Madeleine Hernandez, Brandi Weaber, Andrea Gonzalez, and Sarey Martin Concepción were tremendously helpful in preparing the manuscript and providing editorial feedback, and Rebecca Sok also played a key role in the writing and administration of the grant with the assistance of Usha Stewart, for which we are truly grateful. Kutter Callaway, Adam Green, Michael Murray, and several anonymous referees were kind enough to read through and comment on the entire manuscript. Finally, we thank Jim Stump and Kathryn Applegate for editorial advice and directing us to InterVarsity Press, and we greatly appreciate the patient guidance of David Congdon, Jon Boyd, and the team at IVP.

Madeleine Hernandez wrote a brief chapter-by-chapter study and discussion guide based on an early draft of the book. We have revised it and appended it for personal or small group use. We hope you find it helpful.

STUDY GUIDE

Madeleine Hernandez *and* Justin Barrett

Introduction and Chapter One: Wrestling With Evolutionary Psychology, Embracing Christian Theology

1. The authors claim to be writing to two audiences: people who are convinced of the authority of the Bible (properly interpreted) but not so convinced that humans have been created by God through a process of evolution from other organisms, and those convinced by the case for human evolution but not confident that such a view can fit with a traditional, high view of the Bible's authority. Which audience do you more closely identify with? Why so?

2. Justin Barrett shares his personal struggles with accepting both evolution generally and evolutionary psychology specifically. Which aspects of his story do you resonate with? Are there impressions that you have of evolutionary psychology that seem particularly attractive to you or particularly upsetting based on what you know now?

3. Chapter one includes some general theological concepts. Which of these are familiar to you from your religious tradition? Some of these ideas are revisited elsewhere in the book, especially chapters seven and eight. Which ideas would like to know more about?

4. Have you ever jumped off the edge of a pool or a deck to try to land on a raft or similar object floating in the water? Can you recall a similar situation you have experienced in which overcoming an obstacle or challenge changed the situation for others who came after?

Chapter Two: What Is Thriving?

1. Before reading chapter two, how did you personally define thriving? Did you consider yourself to be thriving? What about in relation to how the authors define thriving?
2. Consider the briefly discussed "right-now" and "not-yet-but-becoming" dimensions of thriving as they pertain to yourself at this moment in your life. Can you recognize instances in your past that could have been part of your not-yet-but-becoming as it relates to your thriving today? Can you also recognize not-yet-but-becoming dimensions of who you are today as they relate to who you might be in the future?
3. Identify additional maturationally natural abilities according to Robert McCauley's guidelines. Are there skills and abilities that, though not fully natural, would cause concern if absent in a human? Can reading and writing be considered part of human nature? Consider ancient cultures where communication occurred through cave paintings or simplified shapes and others where oral traditions and histories were emphasized.
4. The authors state that thriving "is partly an expression of nature and partly the fit of that nature within a niche." Identify three living things (not already discussed in this chapter) and explain how they thrive both from a naturalistic, evolutionary perspective and from a theologically based telos perspective. Are you drawn to one perspective more than another? Why?
5. When it comes to niche construction, we humans are the most extreme. Discuss what niche construction was like for your grandparents and parents and compare accounts with others in your discussion group, paying attention to differences in cultures, ages, and geographic locations. Then, discuss what niche construction is like for you. What do you think niche construction will be like for your children and the next generation? Consider the impacts of technology, climate change, and global politics.

6. Take some time to hypothesize ways human nature can be used to close the nature-niche gap. Do you think such efforts are possible or futile? Why or why not?

Chapter Three: But Why Aren't We Thriving? Nature-Niche Gap

1. How do fitness and human nature fit together to create a holistic view of our species? How are the two concepts the same? How are they different?
2. If not already familiar, take a brief moment to do an internet search on the field of evolutionary psychology. What do you notice about those who negatively view this particular field? What is their main argument? Do you agree with them? Why or why not?
3. Read through the authors' discussion of the three categories and the features within them that make humans unique and identify how each of them contributes to the uniquely human telos.
4. The authors state that their list of features that make humans distinct is not necessarily exhaustive. What traits or categories of features are missing from this discussion? Are there capacities or categories that are discussed that should not have been included?
5. Without reading the next few chapters, hypothesize how "distinctive human thriving" can be achieved through the "proper deployment" of distinctive human features.
6. Identify additional ways our Stone Age minds fail to address our present needs.
7. The concept of morality has long been contested by scholars. In the presentation here, did you hear anything you had not considered before in relation to morality and where it comes from? What, if anything, seems to be missing at this point in the book?
8. Consider that the idea of a human purpose outside of vocation or family place is a rather recent development, thanks to the human nature-niche gap. Discuss what your purpose would be during the

time of your grandparents and great-grandparents. Then postulate the future: what kind of purpose will your descendants search for? Will the gap be closed, and, if not, what new existential struggles might future generations endure?

Chapter Four: Social Gaps

1. Think about everything you did in the last three hours—where you went, what you did, who you interacted with, and so on. Then identify every seemingly automatic, mundane mental task your distinctively social brain accomplished without any apparent effort, such as gathering information about those in your social circle, mindreading, and communicating with language or actions.
2. Take a few moments to think about times when your mind-reading or mentalizing failed: you couldn't tell what someone was thinking or feeling, and it was clear that this failure was inhibiting a harmonious interaction. Now imagine never being able to know what anyone else was thinking, feeling, or paying attention to. How does that make you feel? How do you think others would perceive you? Do you know anyone who has this personal experience?
3. In research on social networks, participants often try to list as many people as they can with whom they have personal relationships and then rate people on this list for emotional closeness and the frequency with which they initiate interaction with them. The result of such an exercise is usually several "rings" of social closeness: around five, fifteen, fifty, and one hundred fifty, but the exact numbers vary. Try this exercise. List who is in your inner ring. Who do you feel like you have a high degree of trust and intimacy with? Who would you readily loan money to? Who would you drop everything for if they were in need? Who would you turn to in a moment of private crisis? Now list who is in the next ring: those who may not be your first go-to people but you still would feel

comfortable asking for a big favor. Do you notice any patterns concerning who is in those rings and how long they have been there?

4. Discuss how the internet and social media have affected social intelligence, emotions, and bonding. Consider things like online dating, anonymous comments, and trolling.
5. It's no secret that the internet's anonymous commenting culture has led to the view that humanity's ability to empathize is lacking. In a world where cities are only getting bigger, how can human nature be used to bridge this nature-niche gap?
6. As the authors state, modern-day nuclear families may not typically live close to their extended families, which would allow adolescents naturally to be invested in by the adults around them. Identify five adults you know who might be able to invest in adolescents today and how parents and adolescents might go about securing such relationships.

Chapter Five: Getting and Using Information

1. Take a moment to think of five individuals you do not know personally whom you would consider "inspirational" or "heroic." Why do you find them so? Have you found yourself purchasing any products they recommend or promote or taking part in health and wellness practices they recommend? Why or why not?
2. Consider the reputational fallout that occurs when a loved or prestigious celebrity of any kind is found to be guilty of "human" behavior ("human" here refers to the imperfect human nature). Does the public hold celebrities to a higher moral bar? Why or why not?
3. The authors state that if a content bias "resonates with a Stone Age mind, it will be more attractive to contemporary humans." Identify three examples of this not mentioned in the text.
4. Using the information the authors present regarding the educational niche gap, discuss how an elementary, middle, or high school could more fully address that gap for children and adolescents.

5. Is there something akin to the educational niche gap for adults? Why or why not? If so, where can this be observed, and how can our human nature be used to address this gap?
6. Identify a few ways adults can guide adolescents in their search for their spark. Was there an adult who was an important part of your own path toward your spark?

Chapter Six: Self-Control

1. We usually think of self-control as an unqualified virtue, but there are times when exercising it can lead to unhealthy or problematic behavior. Identify a few such behaviors that arise from self-control. What are the morals and values behind these behaviors?
2. Consider self-control alongside the ubiquity of technologies that keep us constantly connected to the internet. Is self-control helped or hindered by such things? Provide examples and support from any previous chapter.
3. Discuss the nuances between self-control and discipline. Where do the two converge and where do they separate?
4. Is habitual self-control (i.e., discipline) enough to span the authors' perceived self-control gap? If not, suggest additional measures.
5. What are some ways we modern humans can use our nature to bridge the emotional regulation gap? What are some ways we can use our niche to do so?
6. Haidt's moral foundation theory posits that much of our moral thinking comes from evolved mechanisms that arose because of their ability to improve human fitness. Is the idea that our moral reasoning comes from natural intuitions compatible with Christian views on morality? Consider moral foundations theory in light of Romans 1. Does Haidt's position challenge or support the idea that there are real moral truths?

7. Identify a few ways moral foundations theory can address the self-control and morality gap resulting from a decline in religion.

Chapter Seven: A Summary and a Puzzle

1. Think about a time when you experienced an "ought" that seemed to bump up against a "can." Next, think about a time when you have been tempted to think something is the way it ought to be because that's naturally how it is.
2. Consider normatives ("should," "ought") and human nature. What are some normatives that are automated or implied within our human nature?
3. The authors discuss the convergence of theological anthropology and evolutionary psychology as seen in aspects of human nature. What does this kind of observation mean for both science and religion? Can such a convergence between the two fields be seen elsewhere?
4. Do you think it is possible for both the property and functional approaches to the image of God to be right? Can this be said to be the case for the "underlying human nature" that Wolterstorff refers to? How does his conclusion relate to human thriving? What does it mean for humanity?
5. In chapters three and six, the authors discuss morality and moral foundations theory. What is the relationship between these ideas and the authors' statement that "Christian theology tells us" what we should do with our human nature and informs our telos?
6. Using the facets of human nature that evolutionary psychology has laid out for us thus far and the information presented in this chapter, discuss how religion is (or is not) part of human nature. How does religion address the nature-niche gap?
7. The authors state that Christian theology gives the capacities of human nature described by evolutionary psychology a direction in terms of human thriving. Do you agree with this? Why or why not?

If not, what do you think informs us concerning how we use our nature to address the nature-niche gap? What do you think gives us a purpose?

Chapter Eight: What Is Your Telos?

1. Consider the act of loving others and extending your moral circle within the context of our current sociopolitical climate using the scientific and theological frameworks provided by the authors in this chapter. Provide real-life examples in your discussion.
2. Where is the line between sin nature and the science of human nature?
3. What are some religious and nonreligious obstacles that keep science and the discussion of God separate? What can bridge them?
4. Identify some common arguments and ideologies that prevent humans, especially Christians, from seeing humanity, nature, and human creations as part of God's creation and worthy of care.
5. The authors discuss the general purpose of families. Consider a similar discussion of extended multigenerational families. Do the same principles apply or do they need adjustment? What about families of two—couples? How does this analysis apply to couples who cannot or choose not to bear children (for health, economic, or other reasons)?
6. What modern sources of telos discovery exist today? Discuss their strengths and weaknesses.
7. Should any changes be made to the telos questions offered by the authors at the end of the chapter? Are there any additional questions that should be asked?
8. Are there portions of this book you think could be integrated into other learning communities (churches, schools, etc.)?
9. What (if any) practical steps might you engage in as a result of reading this book?

REFERENCES

Aiello, L. C., & Wheeler, P. (1995). The expensive-tissue hypothesis: The brain and the digestive system in human and primate evolution. *Current Anthropology, 36*, 199-221.

Applegate, K., & Stump, J. (Eds.). (2016). *How I changed my mind about evolution: Evangelical reflections on faith and science*. IVP Academic.

Atran, S. (2002). *In gods we trust: The evolutionary landscape of religion*. Oxford University Press.

Aquinas, *T. Summa Contra Gentiles*, III. 38.

Aquinas, *T. Summa Theologiae*, I, q. 2 a. 1, ad 1.

Balswick, J. O., King, P. E., & Reimer, K. S. (2016). *The reciprocating self: Human development in theological perspective* (2nd ed.). IVP Academic.

Baron-Cohen, S. (1995). *Mindblindness: An essay on autism and theory of mind*. MIT Press.

Barrett, J. L. (2011). *Cognitive science, religion, and theology: From human minds to divine minds*. Templeton Press.

Barrett, J. L. (2012). *Born believers: The science of children's religious beliefs*. Free Press.

Barrett, L., Dunbar, R. I. M., & Lycett, J. E. (2002). *Human evolutionary psychology*. Palgrave-Macmillan/Princeton University Press.

Barrett, J. L., & Greenway, T. S. (2017). *Imago Dei* and animal domestication: Cognitive-evolutionary perspectives on human uniqueness. In C. Lilley & D. Pedersen (Eds.), *Human origins and the image of God: Essays in honor of J. Wentzel van Huyssteen* (pp. 64-81). Eerdmans.

Barrett, J. L., & Jarvinen, M. J. (2015). Evolutionary byproducts and *Imago Dei*. In M. Jeeves (Ed.), *The emergence of personhood: A quantum leap?* (pp. 163-83). Eerdmans.

Barrett, J. L., & Keil, F. C. (1996). Anthropomorphism and God concepts: Conceptualizing a non-natural entity. *Cognitive Psychology, 31*, 219-47.

Barth, F. (1975). *Ritual and knowledge among the Baktaman of New Guinea.* Yale University Press.

Baumeister, R. (2012, July 24). Can virtuous habits be cultivated? *Big Questions Online.* www.bigquestionsonline.com/2012/07/24/can-virtuous-habits-cultivated/

Baumeister, R. F., Vohs, K. D., & Tice, D. M. (2007). The strength model of self-control. *Current Directions in Psychological Science, 16*, 351-55. https://doi.org/10.1111/j.1467-8721.2007.00534.x

Bavinck, H. (2011). *Reformed dogmatics: Abridged in one volume.* J. Bolt (Ed.) (p. 284). Baker Academic.

Benson, P. L. (2006). *All kids are our kids: What communities must do to raise caring and responsible children and adolescents* (2nd ed.). Jossey Bass.

Benson, P. L. (2008). *Sparks: How parents can help ignite the hidden strengths of teenagers.* Jossey Bass.

Benson, P. L., & Scales, P. C. (2009). The definition and preliminary measurement of thriving in adolescents. *Journal of Positive Psychology, 4*, 85-104. https://doi.org/10.1080/17439760802399240

Bering, J. M. (2011). *The belief instinct: The psychology of souls, destiny, and the meaning of life.* W. W. Norton & Company.

Bowlby, J. (1982). *Attachment and loss: Vol. 1. Attachment* (2nd ed.). Basic Books.

Boyd, R. (2018). *A different kind of animal: How culture transformed our species.* Princeton University Press.

Boyer, P. (2001). *Religion explained: The evolutionary origins of religious thought.* Basic Books.

Boyer, P., & Liénard, P. (2006). Why ritualized behavior? Precaution systems and action parsing in developmental, pathological and cultural rituals. *Behavioral and Brain Sciences, 29*, 595-613.

Bundick, M. J., Yeager, D. S., King, P. E., & Damon, W. (2010). Thriving across the life span. In W. F. Overton & R. M. Lerner (Eds.), *The handbook of life-span development: Vol. 1. Cognition, biology, and methods* (pp. 882-923). John Wiley & Sons Inc.

Burdett, M. (2020). Niche construction and the functional model of the image of God. *Philosophy, Theology and the Sciences*, 7, 158-180. https://doi.org/10.1628/ptsc-2020-0015

Cairó, O. (2011). External measures of cognition. *Frontiers in Human Neuroscience, 5*, 108. https://doi.org/10.3389/fnhum.2011.00108

Calvin, Jean. *The Institutes*, I.3.

Christensen, K. M., Hagler, M. A., Stams, G. J., Raposa, E. B., Burton, S., & Rhodes, J. E. (2020). Non-specific versus targeted approaches to youth mentoring: A follow-up meta-analysis. *Journal of Youth and Adolescence, 49*(5), 959-72. https://doi.org/10.1007/s10964-020-01233-x

Clark, K. J., & Barrett, J. L. (2010). Reformed epistemology and the cognitive science of religion. *Faith and Philosophy, 27*(2), 174-89. https://doi.org/10.5840/faithphil201027216

Cohen, E., Ejsmond-Frey, R., Knight, N., & Dunbar, R. I. M. (2010). Rowers' high: Behavioural synchrony is correlated with elevated pain thresholds. *Biology Letters, 6*(1), 106-8. https://doi:10.1098/rsbl.2009.0670

Cosmides, L. (1989). The logic of social exchange: Has natural selection shaped how humans reason? Studies with the Wason selection task. *Cognition, 31*(3), 187-276.

Cosmides, L., & Tooby, J. (1997). *Evolutionary psychology: A primer.* Center for Evolutionary Psychology, University of California Santa Barbara. www.cep.ucsb.edu/primer.html

Curry, O. S., Chesters, M. J., & Van Lissa, C. J. (2019). Mapping morality with a compass: Testing the theory of "morality-as-cooperation" with a new questionnaire. *Journal of Research in Personality, 78*, 106-24. https://doi.org/10.1016/j.jrp.2018.10.008

Curry, O. S., Mullins, D. A., & Whitehouse, H. (2019). Is it good to cooperate? Testing the theory of morality-as-cooperation in 60 societies. *Current Anthropology, 60*(1), 47-69. https://doi.org/10.1086/701478

Damon, W. (1997). *Youth charter: How communities can work together to raise standards for our children.* Free Press.

Damon, W. (2008). *Path to purpose: Helping our children find their calling in life.* Free Press.

Damon, W., & Colby, A. (2015). *The power of ideals: The real story of moral choice.* Oxford University Press.

Diamond, J. (2003). *Guns, germs, and steel: The fates of human societies* (2nd ed.). W. W. Norton & Company.

Diener, E. (1979). Deindividuation, self-awareness, and disinhibition. *Journal of Personality and Social Psychology, 37*(7), 1160-71. https://doi.org/10.1037/0022-3514.37.7.1160

Duckworth, A., & Gross, J. J. (2014). Self-control and grit: Related by separable determinants of success. *Current Directions in Psychological Science, 23*, 319-25. https://doi.org/10.1177/0963721414541462

Dunbar, R. I. M. (1993). The co-evolution of neocortical size, group size and language in humans. *Behavioral and Brain Sciences, 16*(4), 681-735.

Dunbar, R. I. M. (1998). *Grooming, gossip, and the evolution of language.* Harvard University Press.

Dunbar, R. I. M. (2004). Gossip in evolutionary perspective. *Review of General Psychology, 8*(2), 100-110. https://doi.org/10.1037/1089-2680.8.2.100

Dunbar, R. I. M. (2009). The social brain hypothesis and its implications for social evolution. *Annals of Human Biology, 36*(5), 562-72. https://doi.org/10.1080/03014460902960289

Dunbar, R., Barrett, L., & Lycett, J. (2007). *Evolutionary psychology: A beginner's guide. Human behaviour, evolution, and the mind.* Oneworld.

Esipova, N., Pugliese, A., & Ray, J. (2013, May 15). 381 million adults worldwide migrate within countries. *Gallup*. https://news.gallup.com/poll/162488/381-million-adults-worldwide-migrate-within-countries.aspx

Falk, D., Zollikofer, C. P. E., Morimoto, N., & Ponce de Leon, M. S. (2012). Metopic suture of Taung (*Australopithecus africanus*) and its implications for hominin brain evolution. *Proceedings of the National Academy of Sciences, 109*(22), 8467-70. https://doi.org/10.1073/pnas.1119752109

Fraley, R. C., & Marks, M. J. (2010). Westermarck, Freud, and the incest taboo: Does familial resemblance activate sexual attraction? *Personality and Social Psychology Bulletin, 36*(9), 1202-12. https://doi.org/10.1177/0146167210377180

Fuentes, A. (2019). *Why we believe: Evolution and the human way of being.* Yale University Press.

Fujimura, M. (2015). *Culture care: Reconnecting with beauty for our common life* (2nd ed.). International Arts Movement and the Fujimura Institute.

Graham, J., Haidt, J., & Nosek, B. (2009). Liberals and conservatives rely on different sets of moral foundations. *Journal of Personality and Social Psychology, 96*, 1029-46.

Granqvist, P., Mikulincer, M., Gewirtz, V., & Shaver, P. R. (2012). Experimental findings on God as an attachment figure: Normative processes and moderating effects of internal working models. *Journal of Personality and Social Psychology, 103*, 804-18. https://doi.org/10.1037/a0029344

Green, A. (2013). Cognitive science and the natural knowledge of God. *The Monist, 96*, 399-419.

Greenway, T. S., & Barrett, J. L. (2018). Cognitive science, *sensus divinitatis*, and Christ. In A. Torrance & T. McCall (Eds.), *Christ and the created order* (pp. 239-52). Zondervan Academic.

Greenway, T. S., Barrett, J. L., & Furrow, J. L. (2016). Theology and thriving: Teleological considerations based on the doctrines of Christology and soteriology. *Journal of Psychology and Theology, 44*(3), 179-89.

Grenz, S. J. (2001). *The social god and the relational self.* Westminster John Knox.

Haidt, J., & Joseph, C. (2004). Intuitive ethics: How innately prepared intuitions generate culturally variable virtues. *Daedalus Journal of the American Academy of Arts & Sciences, 133*(4), 55-66. https://doi.org/10.1162/0011526042365555

Haidt, J., & Joseph, C. (2007). The moral mind: How five sets of innate intuitions guide the development of many culture-specific virtues, and perhaps even modules. In P. Carruthers, S. Laurence, & S. Stich (Eds.), *The innate mind* (Vol. 3) (pp. 367-92). Oxford University Press.

Haidt, J., Koller, S. H., & Dias, M. G. (1993). Affect, culture, and morality, or is it wrong to eat your dog? *Journal of Personal and Social Psychology, 65*, 613-28.

Hamilton, W. D. (1964). The genetical evolution of social behaviour. I. *Journal of Theoretical Biology*, 7(1), 1-16. https://doi.org/10.1016/0022-5193(64)90038-4

Henrich, J. (2015). *The secret of our success: How culture is driving human evolution, domesticating our species, and making us smarter*. Princeton University Press.

Henrich, J., & Broesch, J. (2011). On the nature of cultural transmission networks: Evidence from Fijian villages for adaptive learning biases. *Philosophical Transactions of the Royal Society B, 366*, 1139-48. https://doi.org/10.1098/rstb.2010.0323

Henrich, J., & Gil-White, F. (2001). The evolution of prestige: Freely conferred deference as a mechanism for enhancing the benefits of cultural transmission. *Evolution and Human Behavior*, 22(3), 165-96.

Henrich, J., & McElreath, R. (2007). Dual inheritance theory: The evolution of human cultural capacities and cultural evolution. In R. Dunbar & L. Barrett (Eds.), *Oxford handbook of evolutionary psychology* (pp. 555-70). Oxford University Press.

Henshilwood, C., & D'Errico, F. (Eds.). (2011). *Homo symbolicus: The dawn of language, imagination, and spirituality*. John Benjamins Publishing Company.

Hill, R. A., & Dunbar, R. I. M. (2003). Social network size in humans. *Human Nature, 14*(1), 53-72.

Hornbeck, R. G. (2013). *Pure war: Moral cognition and spiritual experiences in Chinese World of Warcraft* [Unpublished doctoral dissertation]. University of Oxford.

Järnefelt, E., Ford Canfield, C., & Kelemen, D. (2015). Divided mind of a disbeliever: Intuitive beliefs about nature as purposefully created among different groups of non -religious adults. *Cognition, 140*, 72-88. https://doi.org/10.1016/j.cognition.2015.02.005

Jeeves, Malcolm. (2015). *The emergence of personhood: A quantum leap?* Eerdmans.

Jessen, S., & Grossmann, T. (2014). Unconscious discrimination of social cues from eye whites in infants. *Proceedings of the National Academy of Sciences, 111*(45), 16208-13. https://doi.org/10.1073/pnas.1411333111

Jones, J. H. (2011). Primates and the evolution of long, slow life histories. *Current Biology, 21*(18), 708-17. https://doi.org/10.1016/j.cub.2011.08.025

Kanwisher, N., McDermott, J., & Chun, M. M. (1997). The fusiform face area: A module in human extrastriate cortex specialized for face perception. *Journal of Neuroscience, 17*(11): 4302-11.

Kelemen, D. (2004). Are children "intuitive theists"?: Reasoning about purpose and design in nature. *Psychological Science, 15*(5), 295-301. https://doi.org/10.1111/j.0956-7976.2004.00672.x

Kelemen, D., & DiYanni, C. (2005). Intuitions about origins: Purpose and intelligent design in children's reasoning about nature. *Journal of Cognition and Development, 6*, 3-31. https://doi.org/10.1207/s15327647jcd0601_2

Kelemen, D., & Rosset, E. (2009). The human function compunction: Teleological explanation in adults. *Cognition, 111*(1), 138-43. https://doi.org/10.1016/j.cognition.2009.01.001

King, P. E. (2016). The reciprocating self: Trinitarian and christological anthropologies of being and becoming. *Journal of Psychology and Christianity, 35*, 215-32.

King, P. E. (2018). Kids and God: Nurturing spirituality and the ability to thrive. In Benjamin D. Espinoza, James Riley Estep, & Shirley Morganthaler (Eds.), *Story, formation and culture*. Wipf & Stock.

King, P. E. (2020). Joy distinguished: Teleological perspectives on joy as a virtue. *Journal of Positive Psychology, 15*, 33-39. https://doi.org/10.1080/17439760.2019.1685578

King, P. E. (in press). Vocation as becoming: Telos, thriving & joy. In David J. Downs, Tina Houston-Armstrong, and Amos Yong (Eds.), *Vocation, formation, and theological education: Interdisciplinary perspectives from Fuller Theological Seminary*. Claremont, CA: Claremont Press.

King, P. E., & Argue, S. (2020). #joyonpurpose: Finding joy on purpose. In D. White and S. Farmer (Eds.), *Joy as guide to youth ministry*. General Board of Higher Education and Ministry of the United Methodist Church.

King, P. E., Barrett, J., Greenway, T., Schnitker, S. A., & Furrow, J. L. (2018). Mind the gap: Evolutionary psychological perspectives on human thriving. *Journal of Positive Psychology, 13*(4), 336-45. https://doi.org/10.1080/17439760.2017.1291855

King, P. E., & Defoy, F. (2020). Joy as a virtue: The means and ends of joy. *Journal of Psychology and Theology, 48*(4), 308-31. https://doi.org/10.1177/0091647120907994

King, P. E., & Mangan, S. (in press). Hindsight in 2020: Looking back and forward to positive youth development and thriving. In L. Crockett, G. Carlo, & J. Schulenberg (Eds.), *APA handbook of adolescent and young adult development*. Washington, DC, US: American Psychological Association.

King, P. E., & Whitney, W. (2015). What's the "positive" in positive psychology? Teleological considerations based on creation and imago doctrines. *Journal of Psychology and Theology, 43*(1), 47-59. https://doi.org/10.1177/009164711504300105

King, P. E., Schnitker, S. A., & Houltberg, B. (2020). Religious groups and institutions as a context for moral development: Religion as fertile ground. In L. A. Jensen (Ed.), *The Oxford handbook of moral development*. New York: Oxford University Press.

Kirkpatrick, L. A. (2005). *Attachment, evolution, and the psychology of religion*. Guilford Press.

Kinzler, K. D., Shutts, K., De Jesus, J., & Spelke, E. S. (2009). Accent trumps race in guiding children's social preferences. *Social Cognition, 27*(4), 623-34. https://doi.org/10.1521/soco.2009.27.4.623

Kobayashi, H., & Kohshima, S. (2001). Unique morphology of the human eye and its adaptive meaning: Comparative studies on external morphology of the primate eye. *Journal of Human Evolution, 40*(5), 419-35. https://doi.org/10.1006/jhev.2001.0468

Kurzban, R., Tooby, J., & Cosmides, L. (2001). Can race be erased? Coalitional computation and social categorization. *Proceedings of the National Academy of Sciences, 98*(26), 15387-92. https://doi.org/10.1073/pnas.251541498

Laland, K. N. (2017). *Darwin's unfinished symphony: How culture made the human mind*. Princeton University Press.

Laland, K. N., Odling-Smee, J., & Feldman, M. (2000). Niche construction, biological evolution and cultural change. *Behavioral and Brain Sciences, 23*, 131-75.

Latané, B., & Darley, J. M. (1969). Bystander "apathy." *American Scientist, 57*(2), 244-68.

Lerner, R. (2007). *The good teen: Rescuing adolescents from the myth of storm and stress*. Random House.

Lerner, R. M., Lerner, J. V., Bowers, E., & Geldhof, J. G. (2015). Positive youth development and relational-developmental-systems. In W. F. Overton & P. C. Molenaar (Eds.), R. M. Lerner (Editor-in-chief), *Handbook of child psychology and developmental science: Vol. 1. Theory and method* (7th ed.) (pp. 607-51). Wiley.

Lewis, C. S. (1943). *The Abolition of Man*. Macmillan.

Lewis, C. S. (1960). *The Four Loves*. Geoffrey Bles.

McCauley, R. N. (2011). *Why religion is natural and science is not*. Oxford University Press.

McCullough, M. E., & Willoughby, B. L. B. (2009). Religion, self-regulation, and self-control: Associations, explanations, and implications. *Psychological Bulletin, 135*, 69-93.

McMurdie, C. A., Dollahite, D. C., & Hardy, S. A. (2013). Adolescent and parent perceptions of the influence of religious belief and practice. *Journal of Psychology and Christianity, 32*(3), 192-205.

Meltzoff, A. N., & Moore, M. K. (1983). Newborn infants imitate adult facial gestures. *Child Development, 54*(3), 702-9.

Moll, H., Richter, N., Carpenter, M., & Tomasello, M. (2008). Fourteen-month-olds know what "we" have shared in a special way. *Infancy 13*(1), 90-101.

Mouw, Richard J. (2012). The *imago Dei* and philosophical anthropology. *Christian Scholars Review, 41*(3), 253-66.

Murray, M. J. (2019). Reverse engineering the *imago Dei*. Unpublished paper presented at the Toronto Christian Scholar Symposium, Wycliffe College, January 25, 2019.

Norenzayan, A. (2013) *Big gods: How religion transformed cooperation and conflict.* Princeton University Press.

Öhman, A., & Mineka, S. (2001). Fears, phobias, and preparedness: Toward an evolved module of fear and fear learning. *Psychological Review, 108*(3), 483-522. https://doi.org/10.1037/0033-295X.108.3.483

Olatunji, B. O., Haidt, J., McKay, D., & David, B. (2008). Core, animal reminder, and contamination disgust: Three kinds of disgust with distinct personality, behavioral, physiological, and clinical correlates. *Journal of Research in Personality, 42,* 1243-59.

Pew Forum on Religion and Public Life. (2008, February). *U.S. religious landscape survey: Religious affiliation: Diverse and dynamic.* https://assets.pewresearch.org/wp-content/uploads/sites/11/2013/05/report-religious-landscape-study-full.pdf

Pew Forum on Religion and Public Life. (2012, October 9). "Nones" on the rise: One-in-five adults have no religious affiliation. www.pewresearch.org/wp-content/uploads/sites/7/2012/10/NonesOnTheRise-full.pdf

Raposa, E. B., Ben, E. A., Olsho, L. E. W., & Rhodes, J. (2019). Birds of a feather: Is matching based on shared interests and characteristics associated with longer youth mentoring relationships? *Journal of Community Psychology, 47*(2), 385-97.

Rehm, J., Steinleiner, M., & Lilli, W. (1987). Wearing uniforms and aggression: A field experiment. *European Journal of Social Psychology, 17,* 357-60. https://doi.org/10.1002/ejsp.2420170310

Renner, B. (2018, October 31). Study: Anxiety, depression, panic attacks all on rise among U.S. college students. *Study Finds.* www.studyfinds.org/study-mental-health-anxiety-college-students-rising

Romer, D., Duckworth, A. L., Sznitman, S., & Park, S. (2010). Can adolescents learn self-control? Delay of gratification in the development of control over risk taking. *Prevention Science, 11,* 319-30. https://doi.org/10.1007/s11121-010-0171-8

Rosenberg, S. P., Burdett, M., Lloyd, M., and Van den toren, B. (Eds.). (2018). *Finding ourselves after Darwin: Conversations on the image of God, original sin, and the problem of evil.* Baker.

Roser, M., & Ortiz-Ospina, E. (2019). *Global extreme poverty.* Our World in Data. https://ourworldindata.org/extreme-poverty

Roth, G., & Dicke, U. (2005). Evolution of the brain and intelligence. *Trends in Cognitive Science, 9*(5), 250-57. https://doi.org/10.1016/j.tics.2005.03.005

Rozin, P., Millman, L., & Nemeroff, C. (1986). Operation of the laws of sympathetic magic in disgust and other domains. *Journal of Personality and Social Psychology, 50*(4), 703-12. http://dx.doi.org/10.1037/0022-3514.50.4.703

Sahlins, M. (1972). *Stone Age Economics.* Aldine Publishing.

Santos, A. (2019, September 13). *Why the world is becoming more allergic to food.* BBC. www.bbc.com/news/health-46302780

Santos, R. G., Chartier, M. J., Whalen, J. C., Chateau, D., & Boyd, L. (2011). Effectiveness of school-based violence prevention for children and youth: Cluster randomized field trial of the Roots of Empathy program with replication and three-year follow-up. *Healthcare Quarterly, 14,* 80-91.

Scales, P., Benson, P., & Roehlkepartain, G. (2011). Adolescent thriving: The role of sparks, relationships, and empowerment. *Journal of Youth and Adolescence, 40*(3), 263-77.

Schnitker, S. A., King, P. E., & Houltberg, B. (2019). Religion, spirituality, and thriving: Transcendent narrative, virtue, and telos. *Journal of Research on Adolescence, 29*(2), 276-90. https://doi.org/10.1111/jora.12443

Shweder, R. A., Mahapatra, M., & Miller, J. G. (1987). Culture and moral development. In J. Kagan & S. Lamb (Eds.), *The emergence of morality in young children* (pp. 1-83). University of Chicago Press.

Smith, C. (2003). Religious participation and network closure among American adolescents. *Journal for the Scientific Study of Religion, 42*(2), 259-67. https://doi.org/10.1111/1468-5906.00177

Sosis, R., & Ruffle, B. J. (2003). Religious ritual and cooperation: Testing for a relationship on Israeli religious and secular kibbutzim. *Current Anthropology, 44*(5), 713-22. https://doi.org/10.1086/379260

Sousa, A. L., Byers-Heinlein, K., & Poulin-Dubois, D. (2013). Bilingual and monolingual children prefer native-accented speakers. *Frontiers in Psychology, 4,* 953. https://doi.org/10.3389/fp-syg.2013.00953

Sperber, D. (1996). *Explaining culture: A naturalistic approach.* Blackwell.

Thiselton, A. C. (2015). The image and likeness of God: A theological approach. In M. Jeeves (Ed.), *The emergence of personhood: A quantum leap?* (pp. 184-201). Eerdmans.

Thompson, R. A. (1994). Emotion regulation: A theme in search of a definition. *Monographs of the Society for Research in Child Development, 59*(2/3), 25-52.

Tomasello, M. (2019). *Becoming human: A theory of ontogeny.* Belknap Press.

United Nations Department of Economic and Social Affairs. (2014). *World urbanization prospects: The 2014 Revision, Highlights* (Report No. ST/ESA/SER.A/352). United Nations. https://esa.un.org/unpd/wup/Publications/Files/WUP2014-Highlights.pdf

van Huyssteen, J. W. (2006). *Alone in the world? Human uniqueness in science and theology.* Eerdmans.

Vitz, P. C. (2013). *The faith of the fatherless: The psychology of atheism* (2nd ed.). Ignatius Press.

Volf, M., & Croasmun, M. (2019). *For the life of the world: Theology that makes a difference.* Brazos Press.

Warren, R. (2002). *The purpose driven life.* Zondervan.

Weinstein, D., Launay, J., Pearce, E., Dunbar, R. I. M., & Stewart, L. (2016). Singing and social bonding: Changes in connectivity and pain threshold as a function of group size. *Evolution and Human Behavior, 37*(2), 152-58. https://doi.org/10.1016/j.evolhumbehav.2015.10.002

Wiley, H. (2019, October 14). Abortion pills, gun control and roadkill: New California laws Gavin Newsom just signed. *Sacramento Bee.* www.sacbee.com/news/politics-government/capitol-alert/article236204218.html

Wilkinson, G. S. (1988). Reciprocal altruism in bats and other mammals. *Ethology and Sociobiology, 9*(2-4), 85-100. https://doi.org/10.1016/0162-3095(88)90015-5

Wolterstorff, N. (2008). *Justice: Rights and wrongs.* Princeton University Press.

Whitehouse, H., & Kavanagh, C. M. (2021). What is the role of ritual in binding communities together? In J. Barrett (Ed.), *The Oxford Handbook of Cognitive Science of Religion.* Oxford University Press.

World Health Organization Regional Office for Europe. (2011). *Burden of disease from environmental noise.* World Health Organization/JRC European Commission. www.who.int/quantifying_ehimpacts/publications/e94888.pdf

Xygalatas, D. (2021). Extreme rituals. In J. Barrett (Ed.), *The Oxford handbook of cognitive science of religion.* Oxford University Press.

Yoviene, L., & Rhodes, J. (2016, May 3). There's no substitute for someone who gets you. *Chronicle of Evidence-Based Mentoring.* www.evidencebasedmentoring.org/theres-no-substitute-for-someone-who-gets-you

GENERAL INDEX

SCRIPTURE INDEX

BioLogos Books on Science and Christianity

BioLogos invites the church and the world to see the harmony between science and biblical faith as they present an evolutionary understanding of God's creation. BioLogos Books on Science and Christianity, a partnership between BioLogos and IVP Academic, aims to advance this mission by publishing a range of titles from scholarly monographs to textbooks to personal stories.

The books in this series will have wide appeal among Christian audiences, from nonspecialists to scholars in the field. While the authors address a range of topics on science and faith, they support the view of evolutionary creation, which sees evolution as our current best scientific description of how God brought about the diversity of life on earth. The series authors are faithful Christians and leading scholars in their fields.

www.ivpress.com/academic

biologos.org